ELECTRONIC

A first book on the use, principles and functions of digital computers, presented in a lucid and interesting manner.

It is designed for those interested in computers from a career standpoint and for those concerned with computers in a business context, as well as for the layman who just wants to understand what this comparatively new industry is all about.

TEACH YOURSELF BOOKS

ELECTRONIC COMPUTERS

F. L. Westwater
O.B.E., M.A. (Edin.), M.A. (Cantab.), M.I.E.E.

Revised by David H. Joyce

TEACH YOURSELF BOOKS
Hodder and Stoughton

First printed 1962
Second edition 1967
Third edition 1973
This impression 1976

ISBN 0 340 21435 X

I.B.M. Typesetting by Print Origination Liverpool
Printed in Great Britain for Teach Yourself Books,
Hodder and Stoughton, Ltd,
by Richard Clay (The Chaucer Press) Ltd, Bungay, Suffolk

Contents

Preface to the First Edition

The inclusion of *Electronic Computers* in the 'Teach Yourself' series is a bold step. No book of this kind could do more than enable the layman to understand how a computer works and a little of what it can do. The design, the construction and the use of a modern computer each require the efforts of a large group of people. Even the understanding of a large machine is nowadays the responsibility of more than one designer. It is beyond the scope of a single individual. Fortunately, however, the principles by which a computer functions are not particularly difficult to grasp—it is the sheer physical size, the enormous number of logical elements, and the complexity of organisation of a modern machine of the larger type which puts it beyond the detailed understanding of an individual.

In an attempt to make the subject intelligible without putting too much strain on the ability of the reader to concentrate for long periods, I have described the control system of a wholly imaginary and extremely rudimentary computer which, if it worked at all, would do so only very inefficiently. Nonetheless, if the reader has the patience to follow the description I believe he or she will find the subject begins to make sense.

For the rest, I have described in some detail the various units of which a computer is built up—input and output equipment, storage systems, the arithmetic unit, etc. The applications of computers to some important scientific problems are briefly described.

Although the subject of programming is dealt with in an elementary way in the main text, it was considered desirable to add a special chapter to describe modern methods and recent advances in this subject. This field is becoming specialised, and I have drawn heavily on the expert knowledge of Mr. H. R. Whiteley of the Computer Group, I.C.T. Ltd., for the material in this chapter. I also wish to express my thanks to Miss W. D. Jepps for typing the manuscript.

Finally, it must be made clear that this book deals only with digital computers. Analogue computers operate on different principles and form a quite separate subject.

F.L.W.

1962

Preface to the Third Edition

Computers today are performing prodigious tasks for men. They have guided men to the moon and monitored human performance en route; they produce weather forecasts for one month in advance; they are invaluable to industrial concerns in the production, arrangement, formatting and presentation of information so that an accurate base is provided for decision-making. Progress in the computing world during the last three years has been exceptionally rapid which has meant that large tracts of this work have become obsolete.

Each significantly different innovation in computer technology has been regarded, by the technologists, as a new generation. Computers being now in their third generation (and the fourth seemingly not far off) it seemed appropriate to produce the Third Edition of this work at this time.

D. H. J.
1973

1 Basic Principles

The popular idea of a computer as an electronic 'brain' is not entirely apt. Basically, a computer is merely a calculating machine, which is a natural development of the desk-type calculator with the difference that the speed of calculation has been enormously increased.

The high operating speed of a computer also lends itself to many applications in which computing is only a minor part.

This is because, in addition to being able to calculate, a computer can store very large quantities of information. If we wish to continue the human analogy, we may think of this as a 'memory'. There are a number of applications in which the usefulness lies in the rapidity with which some items of information stored can be brought out or compared with some information just received. This type of activity is nowadays called

Data Processing. We shall give examples of this later.

The actual methods of calculation are very simple; one could almost describe them as moronic. Nevertheless, because of the extremely high speed at which they are carried out, problems formerly considered insoluble by arithmetical means because of the labour required can now be dealt with almost as a matter of routine. A modern computer can perform in a matter of hours a series of arithmetical operations which would occupy a human computer for many months or even years.

In business applications there arises the need for the repetition of a very large number of simple and broadly similar calculations as, for example, in the preparation of a factory payroll. Formerly, this work required a large number of clerks and might take nearly a week in a large organisation. By using a computer of the business type, which can also print out the payslips, the whole job can now be completed in a day or less by a small staff.

Some organisations, such as Life Assurance Companies, maintain records of all their clients, who may number a million. These records may require to be altered and brought up to date and referred to at intervals. By storing the records on reels of magnetic tape all the above operations can be carried out rapidly by a computer.

Before going on in subsequent chapters to describe in some detail how all this is achieved, let us take a look at a computer in a general way so as to see how it does its job.

Electrical Representation of Numbers

The first thing to bear in mind is that we are dealing with an electronic device.

Numbers have to be represented by something that the computer can 'understand', so to speak. This could be, and often is, a voltage level or a magnetic state. At this stage it is simpler to think of a number as being represented in the computer by a group of electrical switches which can be OFF or ON. It is the combination of OFF and ON which represents or codes a number. This is more fully explained in Chapter 2.

The calculating speed of the computer is basically due to the fact that we can set and reset these groups of switches at very high speeds—in one-millionth of a second, for example.

Before it can operate, the computer must have numbers presented in the form which it can understand. This requires some kind of decoder or mechanism which will translate a number into a suitably recognisable electrical impulse. A common way of doing this is first to represent the number physically as a hole in a card or on a length of paper tape. On the card, the position of the hole corresponds to the number required; on a paper tape, holes are punched across the tape according to a predetermined code. The card or tape is then 'read'. This may be done by allowing the holes to pass under tiny wire brushes which enables them to make contact momentarily with a conducting roller and so pass an electrical impulse into the machine.

In other cases the hole or holes are used to allow a beam of light to fall on photocells, which then pass on the electrical impulse. We should mention here that, of course, the information which may be conveyed to the machine by the above methods is not confined to numbers only. By punching holes suitably, letters of the alphabet or other symbols may be transmitted electrically. These are not, of course, used in the calculations, but may be required to operate a printer when the calculation is completed.

The punched-card and paper-tape methods of

supplying information to the computer are in common use and are reliable, but they tend to be regarded as slow in operation for certain purposes.

It is possible to 'write' the information on a reel of magnetic tape of the type used in tape-recording machines. Using magnetic tape, information is stored in much the same way as it is represented inside the computer, that is as a series of ones or zeros (see Chapter 2–The Binary System of Arithmetic). Magnetic tape is a ½-inch wide strip of celluloid coated on one side with magnesium oxide. In much the same way as paper tape the width is 'divided' into seven or nine tracks, each track representing a one or a zero, and the combination of ones and zeros across one 'frame' forms a character, or bit. The character is one when it is magnetised and zero otherwise. Because of the now electromagnetic nature of both reading and writing this information, two things have effect on the reading speed. Firstly, the data may be packed far more tightly on the tape, since there is no longer a physical hole to be punched. A standard writing density is 556 bits (or characters) per inch. Secondly, the mechanical aspects of reading no longer apply, and thus with the increased strength of magnetic tape much faster reel speeds can be obtained. In fact, a nominal rate of transfer of information can be of the order of 100 000 characters per second.

The method of recording, and speed of operation, of this type of media render it relatively impractical as a means of primary input to a computer system. Normally magnetic tape is used for mass storage of information required by the computer during its operation.

Magnetic tape is now subject to international standards, and therefore is the ideal medium in the situation where one would wish for an interchange of information between computer systems.

Stores and Registers

On receiving the raw material for a calculation in the form of figures translated into electrical impulses, the computer acts exactly as a human computer would in the circumstances–it writes the information down. Naturally we use the term 'write' in an electrical sense. What happens is that the impulses are used to hold the information in an electrical store. Numbers, suitably coded, are held in a store until the machine wishes to perform arithmetic operations upon them.

A human computer requires a piece of paper on which to write down the intermediate stages of a calculation. A computer also needs a similar facility, and this is provided by the presence of additional stores which are usually called Registers. There may be several of these, and a normal number of registers would be three.

To take a very simple example, we could transfer two numbers which we wished to add together into two of the registers and put the result of adding them into a third register. If the first two registers were then cleared, two other numbers could be entered, and the calculation could proceed.

Registers in a computer really fulfil the function of the paper on which a human calculator would do his work. Certain results have to be retained for use in the final calculation, and other figures can be rubbed out when they are no longer required.

Registers, then, are working stores, each holding a number consisting of the particular size with which the computer is designed to operate. Thirteen decimal digits is a common size, but machines operating with more or fewer digits exist.

Transfer of numbers between registers is possible, and can be done rapidly.

In addition to these registers in which the arithmetical operations are carried out, it is usually necessary to have a more permanent form of store to hold numerical and perhaps alphabetical information. Stores of this type can be very large and contain up to tens of thousands of numbers or 'words'. We use inverted commas because, of course, all information held in store has to be in a basically numerical code. For convenience a basic number of, say, thirteen decimal digits, to take a common size, is referred to in computer terminology as a 'word', or 'byte'.

Transfer of a word from one part of the store to another is usually possible, but seldom directly. The word has first to come out of the store into a register or equivalent and then be replaced in its new position.

Another requirement is that we must know where to find the word we want in the store. This means that each word has to have an 'address'. By means of electronic switching devices a word is routed to the correct address in the store, and in a similar manner it can be retrieved from any address if required.

We shall stress here that it is the address which is important and not the content. A computer is not interested in the actual value of the numbers but obeys instructions to transfer contents from one address to another address. It is as if a housing dictator issued the order: 'Take the family living in 13 Acacia Avenue and transfer them to 27 West Street'. This operation would take no account of the names or number in the family.

Storage devices will be dealt with more fully in a later chapter, but we should point out now that, apart from capacity, a store has another important characteristic which is referred to as access time. This simply means that it takes a finite time to route information in or out of a large store; this time depends on the design of the store. As usual, speed is an expensive commodity, and frequently a compromise is adopted in which a computer may have a medium-sized fast-access store and a larger store with slower access time.

Comparisons between human and electronic computers should not be pressed too far, but in thinking about stores and registers it may help to compare a fast-access store to the human memory, in which, for example, we hold the multiplication tables. A longer-access-time store might be compared to a book of logarithmic tables which have to be looked up. The relation between access time and size will be realised by anyone who has used the larger type of tables.

A useful application of the storage principle is to what is called a buffer store. This is a device for gaining time. In some cases, in commercial applications, for example, the calculation may take longer than the time available between the passage of successive punched cards. We can solve this problem by reading the card first into a buffer store, thereby making available the time to read a card as well as the time between cards.

In addition to using registers to hold numbers, it is also necessary to have some means of moving the numbers about the system. For example, we may wish to transfer a number from one register to another or to feed it to an arithmetical unit to perform some operation upon it. Numbers can be moved out of a register by a series of electronic pulses, which have the effect of shifting all the numbers in the register one place along the bed, so to speak. Thus the number:

2	6	9	7	5	2	4

is shifted by the first pulse into:

	2	6	9	7	5	2

The number formerly at the right-hand end can either be allowed to pass on to one of the so-called 'highways' in the computer or it can be obliterated, or it is possible to recirculate it by feeding it back in the left-hand end of the register, which then looks as follows:

4	2	6	9	7	5	2

It is usual to have a series of electronic pulses hammering continuously at the register and to admit them when required by opening what is referred to as a 'gate'.

Arithmetical Operations

We must now try to understand how the actual operation of arithmetic is carried out by the computer. Let us suppose that we have two numbers which are held in two registers which we shall refer to as the A and B registers. We wish to add these numbers and put the answer in a third register, the C register.

In the diagram the registers A, B and C are indicated by rectangles, and the circles represent what are called 'gates' because they can be open or

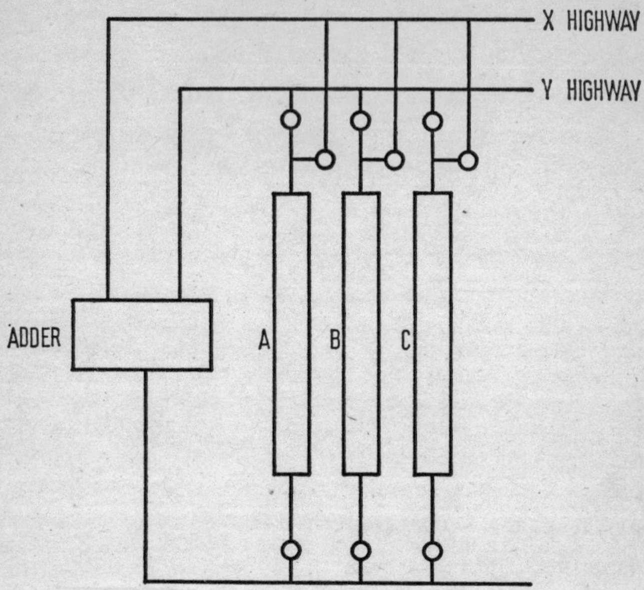

FIG 1.1

shut as required. It will be noticed that these gates can admit or deny access to either of the main highways X and Y. Numbers are driven out on to the highways serially by admitting electronic pulses to the registers in the way we have just described. The adder is an electronic device into which, if we feed two numbers serially, their sum emerges serially. The highways are connected to the adder, and numbers can be transferred from one register to another only by passing through the adder. If only one number is fed into the adder, it comes out unchanged; if two numbers are fed in on the two highways, what emerges is their sum. Numbers coming out of the adder can be admitted to any register by opening the appropriate gate. The operation of subtraction takes the same time as that of addition, since it is reduced to addition by a simple trick which we shall explain later. Multiplication takes several times as long as addition, and division even longer. Even so, the time involved is extremely short by ordinary calculating standards. More complicated calculations, such as finding square roots, can be reduced to some combination of the basic arithmetical operations, and the machine can be instructed to carry these out in turn so as to obtain the correct result.

Machine Functions

The designer of a computer has to arrange that it can carry out certain basic operations of a simple type on command. These functions, as they are called, happen automatically when the correct stimulus is applied. A calculation can be carried out if the necessary functions are carried out in the correct order. The precise order in which they should be carried out is called a program. The individual who, by taking thought and using only the functions supplied by the designer, arranges that the machine can obtain the answer is called a programmer.

For example, a function might be:

Transfer the number in register A to register B.

This could happen if a series of pulses from some source is applied to register A, thus driving the number out of A on to the highway. If at the same time a gate admitting to register B is opened, then the transfer will take place. Thus the performance of the function is reduced to the opening of a series of gates to admit the driving pulses and to permit the number driven out to obtain admittance to the appropriate register. After the correct number of driving pulses has occurred, the process stops if the pulse gate is closed. This will happen if we have a counter to count the pulses, since the size of the register is fixed for a particular machine.

To fix ideas, we can imagine a simple computer with eight functions as follows:

1. Transfer number from Store to A Register.
2. Transfer number from Store to B Register.
3. Add the number in A Register to the number in B Register and put result in C Register.
4. Subtract the number in A Register from the number in B Register and put result in C Register.
5. Test the number in C Register for positive or negative.
6. Test the number in C Register for zero.
7. Transfer the number in C Register to the Store.
8. Stop.

In an actual computer there would have to be some additional functions, but their purpose is not easy to understand at this stage. The example above is given to illustrate the scope which we may expect to find in a simple system.

Programming

Having now seen basically how a computer receives its information and the way in which that information may be processed and manipulated, one should pause before going on to see how a computer processes its information in the appropriate manner. That is to say, if we input to the computer all the relevant information concerning a person's wages, how can we, the user, ensure that the end result is that person's net pay and deductions, accurately calculated in monetary terms from a series of hours, percentages, rates, etc.?

This hypothetical problem surely begs the question 'can the computer think?', and if one were to take at face value the stories, rumours and tales-of-woe which circulate among people not too far removed from 'the machine', then one could easily draw the conclusion that computers can think. However, as is often the case, computers should not be taken at face value. Invariably a fault of 'the computer' should be read as a fault of 'the system' or 'the program', since it is the program which gives the computer direction and it is the system which organises the sequence of programs; and both the system and the programs contained therein are devised, specified, written and tested by highly trained human specialists. It is the conviction of the author that computers do not and, indeed, cannot think; they blindly follow the program's instruction, and if those instructions are erroneous the computer will blandly carry on churning out erroneous and therefore meaningless results.

In order to avoid this situation it is necessary to undergo a lengthy study of the problems before one can finally unleash the computer on calculating the solution. The chain of events starts with a Systems Analyst whose function is to gather together all information concerning the application under review. He has to establish the type and level of input information, the degree, volume and timing of required results, and then decide upon the best method of converting input to output from both a computer and a human point of view. When the Systems Analyst has finished, the resultant document is a System Specification, which describes a series of stages through which our input data must pass in order to produce the desired results. Each stage represents the framework of a program. This document passes to the programmer, who then has to convert the human-readable document into a series of logical steps that the computer can read and carry out.

Computers are essentially numeric machines. They can read numbers and act upon them. In order that some semblance of order can be introduced into their lives a series of instructions, in numeric form, has been devised to ensure this. These instructions form a 'programming language', which in its most basic, numeric form is known as 'machine code'. Not unnaturally it was soon recognised that, although computers were happy ploughing through a set of instructions in numeric form, people were not. So a scheme was devised whereby the numbers were replaced by mnemonics which were of more significance to humans (e.g. MPY = multiply = machine code 064) and a special program was written which enabled the computer to convert the mnemonics to machine code. This conversion process is known as 'compiling' the program.

Each machine code instruction is basically in three parts:

1. The function–what is to be done, e.g. ADX (add into X).
2. The register–where this is to be carried out.
3. The address–that part of store which contains the relevant figure.

Thus from one instruction the computer now has all the information required to process one item of data once only. It is in this way that a programmer sets out to solve the System Specification and convert it into something meaningful to the computer. Let us take a simple example:

Add the information in address 96 to the information in address 97, placing the result in address 100.

Our program would read:

LDX	3	96	this puts the contents of 96 in register 3.
ADX	3	97	adds the contents of 97 to it.
STO	3	100	places the result in 100.
END			tells the computer to stop.

Thus we have the beginnings of a program, and the basis of programming.

However, due to the widening spread of computer applications, it was rapidly established that this form of programming in one-phase steps was too laborious for many people. It took a lot of time and required a high degree of skill and concentration over an extended period.

'If', it was argued, 'compilers can translate one

mnemonic instruction into one machine code instruction, could they not be enhanced to convert one word into a series of standard machine code instructions?' For example, why could one not write the above example as: ADD 96 to 97 GIVING 100, and let the computer translate ADD and GIVING into the necessary machine code.

This gave rise to the advent of 'high level' programming languages which are common to application fields.

The commercial business installations have common functions to perform, so they have devised COBOL (COmmon Business Oriented Language) using key words which a computer recognises and translates, thus enabling programs to be written in a basic form of English.

Scientists wanted to process complicated formulae by computer, so they devised FORTRAN (FORmula TRANslation) which uses accepted standards of formula construction for its conversions.

Mathmeticians work in algorithms, so they now have ALGOL (ALGOrithmetic Language) available to them.

It is the advent of high-level languages such as these which has made possible the spread of computing power and enabled knowledge to come much more rapidly than might have been the case.

One aspect of computer programming which has not been mentioned in this section so far is time. The passage of an electrical impulse round the computer store takes time, and if it has to perform complex calculations during that journey round, then it will take longer. Every programmer is continuously aware of the length of time the operations he is putting together will take. In machine code it is easy to estimate how much time a program will take to run by adding together the instruction times for each instruction in the program. With mnemonics (or Assembler Language) the programmer is one step removed from machine code but can, by clever programming, construct an efficient program. However, when using high-level languages the programmer, during writing, is totally removed from the machine code and is using words which will compile a standard set of instructions, whether the program requires them all or no. Therefore high-level languages tend to produce inefficient programs machine-wise, and a program may well be 'optimised', i.e. rewritten on a time-critical basis, several times during its life.

Presentation of Results

The final requirement is that the computer must be able to present the answer when it has finished the calculation. In business use it must usually present it in some special form. If the answer is the correct figure on a payslip, then the answer must be *printed* on the payslip. It sometimes happens that the answer is required to be punched on a card for use on some subsequent occasion. It is, therefore, necessary to cause the printer or the punch mechanism to operate as a final program instruction. This presents some difficulties, among which may be mentioned the difference in power level required as soon as we pass to the mechanical realm and also the different time scale involved in a mechanical operation. Most of the earlier types of computers displayed the contents of the registers by means of tiny neon lights on the console. Although useful to the programmer and the Maintenance Engineer, the number of lights employed now tends to be less.

2 The Binary System of Arithmetic

The Managing Director drummed impatiently on his desk. Where the devil was that fellow Simpson? It must be a good twenty minutes since he had asked for him to come and explain why a delay in his department looked like holding up production for a week.

There was a discreet tap at the door. 'Mr. Simpson is here now, sir.' Explanations followed. Simpson had been in the sheet-metal shop, where the noise had drowned the loudspeakers asking for him, and a messenger sent to look for him had missed him there.

Afterwards, the Managing Director sent for his personal assistant. 'I'm fed up with this!' he barked. 'Find some method of contacting people in the factory which works. Use flashing lights, pocket radios—anything.'

Young Wilson went thoughtfully away. Lights—

it was certainly an idea. Lights placed high up where everyone could see them, and repeated in each part of the factory where departmental managers were apt to wander. How many people were involved? There were Smith, Jones, Simpson, . . . He counted up to seven, but there might be more later. To spell out the names would be far too elaborate and costly. Some kind of code was what was required. What was wrong with a simple numerical system?

1. Smith   2. Jones   3. Simpson   and so on.

'I may get involved in an argument about seniority,' he mused, 'but we'll have to face that. All I need is a number of boards each mounted with seven lights. When the manager with code number 5 is required we switch on five lights, and so on. Should work like a charm, and won't be too expensive.' He went off to speak to the factory Administrator and told him what he wanted to do.

'It's not a bad idea, my boy, but you don't need seven lamps, three will do, and that will be even cheaper.'

'How do you make that out?'

'Merely by carrying the coding system a bit farther. How's this?' Seizing a sheet of paper, the

Administrator drew a rough sketch (see Fig. 2.1, below).

'You get the idea? Each one of the seven has a unique coding which he can recognise, and as you see, I've done it with only three lamps.'

'Jolly clever, but what happens if we have more than seven people later on?'

'That's quite simple. With four lamps we can make a code for fifteen, and with five lamps for

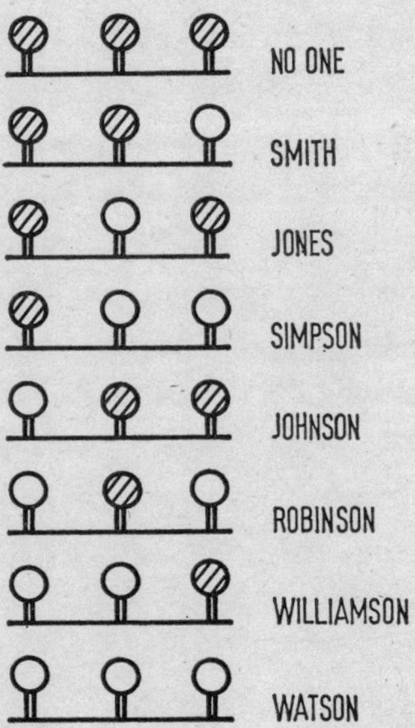

FIG 2.1

thirty-one. What we have really is not a code for Smith, Jones, etc., but a code for ordinary decimal numbers.'

'I think I'm beginning to get the idea,' said Wilson. 'Thanks very much. I'll go and try this on the Old Man.'

We hope this little anecdote will serve as a painless introduction to the binary system of arithmetic, which is fundamental to computers at present. The important feature of the code for identifying the managers was the fact that we had a device which had two distinct states. It was a lamp which could be either ON or OFF. Basically, any other two-state device would have conveyed the same information. For example, two arrows, one pointing up, and the other down. Let us represent the lights as being on or off by the numbers 0 and 1. In this case we could write the code thus:

	Binary	*Decimal*
No one	000	0
Smith	001	1
Jones	010	2
Simpson	011	3
Johnson	100	4

Robinson	101	5
Williamson	110	6
Watson	111	7

There is, of course, no magic about names of people, and we show how the binary code of 0's and 1's could equally well be used to provide a code for the decimal numbers 0 to 7. If we want to continue with a binary code for numbers greater than 7 we have to use a fourth binary 0 or 1, since we have exhausted the possibilities with three. The code for 8 would be 1000, 9 would be 1001, and so on up to 15 (= 1111); after that we need five binary symbols, starting with 16 = 10000. We must now examine this numerical coding system more closely.

Let us start by taking a close look at the system with which we are all familiar–the decimal system. Here we have ten symbols–0, 1, 2, 3, 4, 5, 6, 7, 8, 9. There is no particular merit about the number ten. It is an accident of fate. A kindly Providence gave us ten fingers and, no doubt, primitive man used them to count on, and so ten became the basis of our arithmetical system. We used the word basis, but the term 'radix' is more often used in this context. (It is the Latin word for 'root' and the plural is 'radices'.)

Science has other examples of accidental choice of number units. The fact that we divide a circle into 360 degrees is due to the fact that the early astronomers with their relatively crude instruments thought that there were 360 days in a year. We may be thankful that they did not alter the angular measurement of 365 when the error was discovered!

With ten symbols we can identify up to ten objects. Suppose, however, that we have more than ten? Someone in the dim past made the colossal advance in human thinking in suggesting that we could represent numbers greater than ten by arranging the existing numbers in a definite order.

15	means	10 + 5
27	means	2 x 10 + 7
164	means	100 + 6 x 10 + 4
928	means	9 x 100 + 2 x 10 + 8

and so on. Note that as we move to the left in our arrangement the significance is that each individual digit is to be assumed as being multiplied by a higher power of ten. In this way we can represent very large numbers with the minimum of symbols and, once the rules have been laid down, perform the operations of arithmetic simply and correctly.

Let us now adapt the same principle to a system in which we have only two symbols. We shall take the first two of the decimal system and use these, i.e. our only numbers are 0 and 1. If we want to represent larger numbers we must arrange our 0 and 1 in the same way as the decimal system, but here as we move to the left we ascend in powers of two instead of powers of ten. Thus:

11	means	2 + 1	=	3 in decimal
10	means	2 + 0	=	2 in decimal
101	means	4 + 0 + 1	=	5 in decimal
1110	means	8 + 4 + 2 + 0	=	14 in decimal

and so on. To finish with a larger number we should find that

10111011 means
128 + 0 + 32 + 16 + 8 + 0 + 2 + 1 = 187

'All very interesting,' you may say, 'but what is the point of all this?' The point is simply that the reliable and stable electronic devices are so far *two-state* devices, and so, if we want to represent numbers electronically, we have to use the binary system. If there were electronic systems with ten stable states we could use decimal arithmetic. Such states exist, but they are not sufficiently reliable to use in a computer, although they have found other applications. If we had a three-state device we could easily enough develop a ternary system of arithmetic, but the facts are that the two-state device, e.g. a switch ON or OFF, or a magnet magnetised as N□S or S□N, is reliable and stable and seems unlikely to be displaced for a long time to come. Thus if we want a computer to do arithmetic for us we are forced to arrange that it does it in the binary code, since this is the only reliable code which it can understand. Let us note in passing that all even numbers in the binary system end with a 0, and all odd numbers end with a 1. To multiply by two, we just add a 0 on to the end of the number. This is an exact parallel with the decimal system, where, to multiply by ten, we add a 0 to the number.

With a very little practice, it is quite easy to carry out simple arithmetic in the binary manner. For addition the basic rules are:

1 + 1 = 0 and 1 to carry

1 + 0 = 1

0 + 1 = 1

0 + 0 = 0

Let us try adding three (11) to six (110) in this way:

$$\begin{array}{r} 1\ 1\ 0 \\ 1\ 1 \\ \hline 1\ 0\ 0\ 1 \end{array}$$

and 1001 is, of course, nine as it should be if the sum is correct.

We shall have a little more to say about this in the section on the Arithmetic Unit in Chapter 3.

In many ways, calculations done in the binary manner are simpler than in decimals. Basically this is because there is a multiplication table with only four entries. The difficulty is that we normally start and finish with decimals in an actual problem, so that when we know the answer in binary it is necessary to convert it into decimal form. When dealing with large numbers this is rather a cumbersome process. It does not matter much in a complex mathematical calculation, where the bulk of the time is spent on the actual arithmetic, but it can be a serious disadvantage in commercial applications, where the amount of calculation is small and speed in printing out the answers is the main point of the exercise. It is, of course, possible to arrange for the computer to do the conversion back to decimal for printing out. This requires some storage space and either program steps or some built-in function, and all this is expensive. It is common practice in commercial machines to use a combined code which is a compromise between binary and decimal but in which the arithmetic is done in binary manner. The system is called binary coded decimal, and in this each separate digit in the decimal version is coded in binary. Thus 7359 would be coded 0111, 0011, 0101, 1001. This leads to complications in the arithmetic but has the advantage that the final conversion back to decimals is quite easy to achieve by a simple recode and does not require any storage or arithmetic. Sometimes the binary figures are fed into the computer serially, i.e. as they are written above one after the other; in most cases they are fed in parallel, i.e. each group of four simultaneously. This last method saves time but is more expensive in equipment.

A further complication which the designer of commercial machines has to contend with is that much of his work is not carried out in a simple decimal notation. He has to be able to work in £p and even in tonnes, kilograms, etc.

Electronic Switching

The binary system of notation is the most important development to be utilised in digital computers. As we have already seen, it is developed from a scale of two, as opposed to the scale of ten or unitary system we all must have used in our formative years. The importance lies in the fact that it is comparatively easy to produce reliable electrical or electronic devices capable of switching between two states (i.e. ON/OFF) at speed. These can then be used to represent our two desired binary stable states, or to use a more up-to-date jargon: Flip-Flop or Bistable—Bi, of course, being a commonly used Latin prefix for two.

Various conventions have been used to represent these two states, such as the one already mentioned—ON/OFF, High(Hi)/Low, 0/1, Set/Reset, Up/Down—and any of these can be used to represent a high or low current or voltage state, which can then be used as arithmetic notations for 0 and 1, i.e. 1 = ON = HIGH = Set = Up and 0 = OFF = LOW = Reset = Down.

Examples of suitable switching devices are:

1. Mechanical switch.
2. Relay.

3. Valve, transistor or solid-state device.
4. Magnetic core.

The first two of these examples are used quite widely in the field of production control computers, whilst the latter two are used extensively in our digital computer complexes. Even then, however, some of these are fast becoming historical and the degree of ageing is directly attributable to speed. A mechanical switch can take seconds to traverse between its stable states; a relay, milliseconds; whilst with transistors this can be achieved in microseconds and speeded up to nanoseconds in the case of solid-state devices. (Millisecond = $\frac{1}{1000}$ of a second, microsecond = $\frac{1}{10^6}$ of a second, nanosecond = $\frac{1}{10^9}$ of a second.)

Of course, with present developments in this field, even these speeds will be surpassed in the next decade. Some of the developments you will hear of, e.g. thin-film techniques and fluidics, are all being developed towards the same end, to speed up the function of processing data.

The action of changing the state of our flip-flop or bistable is achieved by use of a small current of voltage pulse often only nanoseconds in duration.

A common symbol for our electronic switch is shown in Fig. 2.2:

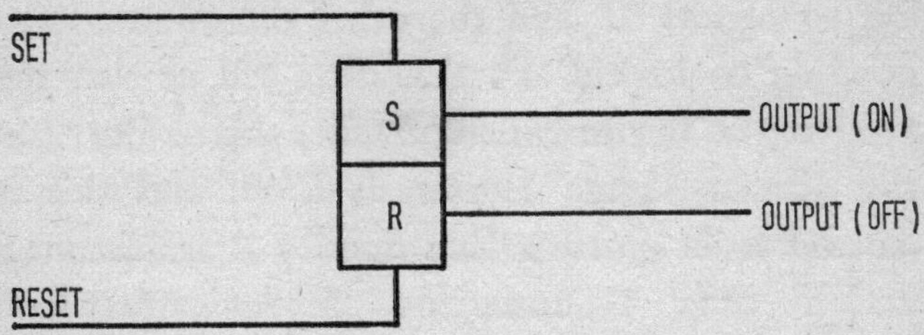

FIG 2.2

It is said that when the bistable is set it is storing 1 or some quantitative amount and will continue to do so until some reset action takes place, and that when reset it stores 0 or some other quantitative amount.

The fourth example mentioned, magnetic cores, will be handled in greater detail in the next section. Sufficient to say at this time that they are used for a more permanent form of storage called 'core storage'.

Magnetic Representation

Ferrite and binary cores are widely used in digital

systems for storage purposes. They are suitable because they may be permanently magnetised in either of two directions, one direction corresponding to logical '0' and the other direction corresponding to logical '1'. Scientists tell us that the property of ferrite which provides this ability 'is in the almost square Hysteresis loop, and this is derived from plotting flux density B against magnetic force H'. To understand this let us first look at the Hysteresis loop produced using a simple electromagnet with a soft iron core (Fig. 2.3).

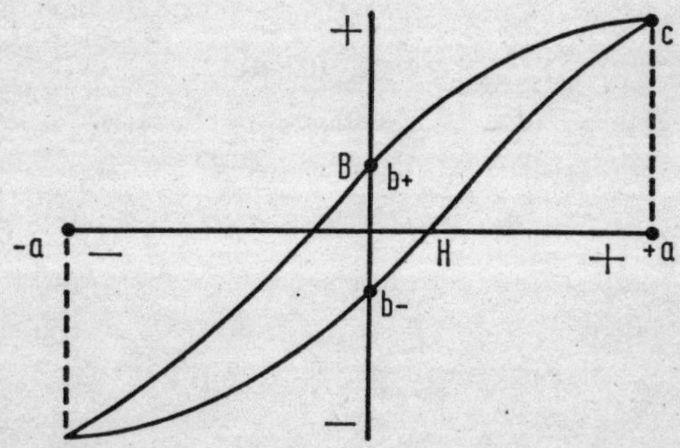

FIG 2.3

The field intensity H in Fig. 2.3 requires to be raised to point *+a* before flux density B is at positive saturation point *c*; then, when the magnetisation force H is removed, the flux density returns to point *b*+, which is a measure of residual magnetism or, as it is called in this context, positive remanence. If we now apply a negative magnetisation force to point −*a*, then the core will be at negative saturation; if we now remove the magnetisation force H, the flux density within the core will drop to *b*−, or negative remanence. Thus we do effectively have two stable states, *b*+ and *b*−. However, because of the core material used, they are of such a small value as to make their use unstable. If we now look at the Hysteresis loop for a piece of ferrite in Fig. 2.4, it must be obvious

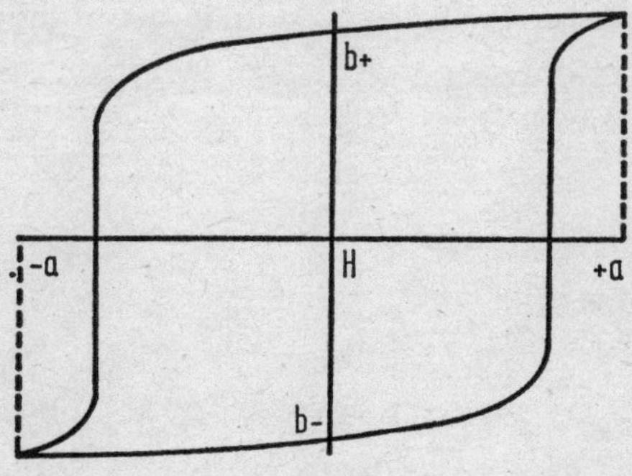

FIG 2.4

that, because of its 'squared' characteristic, the quantities $b+$ and $b-$ are considerably greater than those produced with soft iron. They are, in fact, of sufficient amplitude to be stable in use.

So we now have a device which can store 0 or 1. But how do we utilise this? Well, many systems are used, but by way of explanation we will describe only the Coincident Current Memory System.

There are two main reasons for the selection of the ferrite. First, its two stable states contain much more flux than iron under equal conditions; therefore, when information is read a larger amplitude signal is output as a result. Secondly, a rectangular Hysteresis loop is preferred because it can, by an amplitude selection system, discriminate against small switching currents which are inherent in present-day core storage design.

Coincident Current

Write

Here we show diagrammatically (Fig. 2.5) a core which requires a half-drive current (or half the magnetisation force) down both the X and the Y axis drive windings in the appropriate direction in order to switch the state of the cores and which, having achieved this, will then remain in that state until reverse currents are applied.

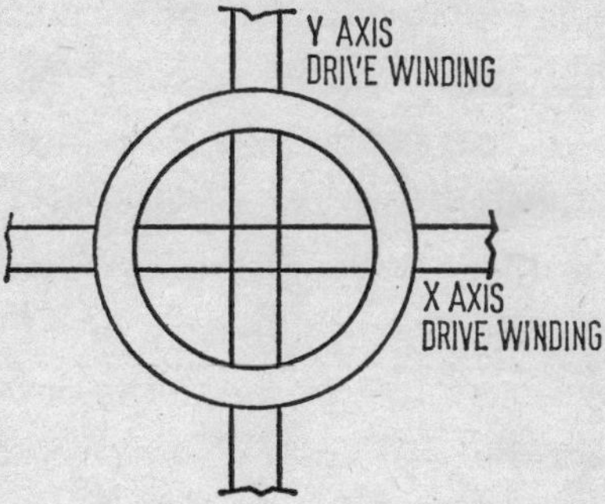

FIG 2.5

Read

In order to read, currents are produced in the X and Y drive windings in the direction necessary to switch the core to its 0 state, and by threading a sense winding through the core the amplitude of the induced current can be used as an indication of the core's state. In Fig. 2.6 below, we show a core selected for read already in the 0 state, and the effect of applying current to switch it to the 0 state will result in driving the flux density to point *c*, which would induce a very small change in the sense winding, i.e. from *b* to *c*; whereas in a core already holding a 1 (Fig. 2.7) the induced effect in the sense winding from driving the core from *b+* to *c* would be large. Therefore, on a core selected for read, a small change in the sense winding would be ignored and a large change taken to indicate that the core had 1 in it. A disadvantage, however, is that, when the magnetising force was removed, then the core would return to negative remanence and hold a 0; so it is said cores are self-destructive on read, and if it is necessary to conserve the information for further actions, it has to be arranged to write back the stored state before modifying it in any way.

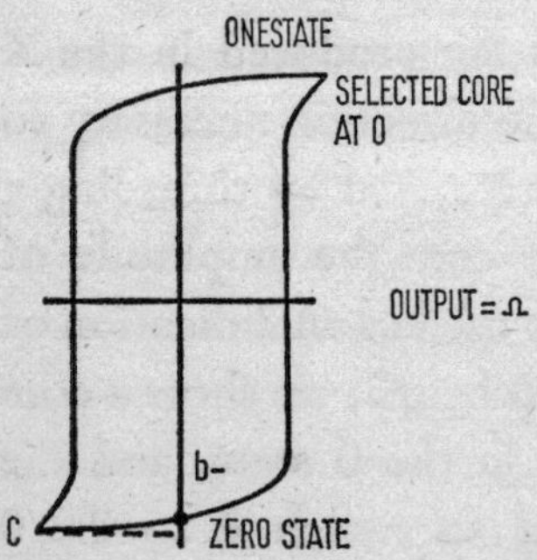

FIG 2.6

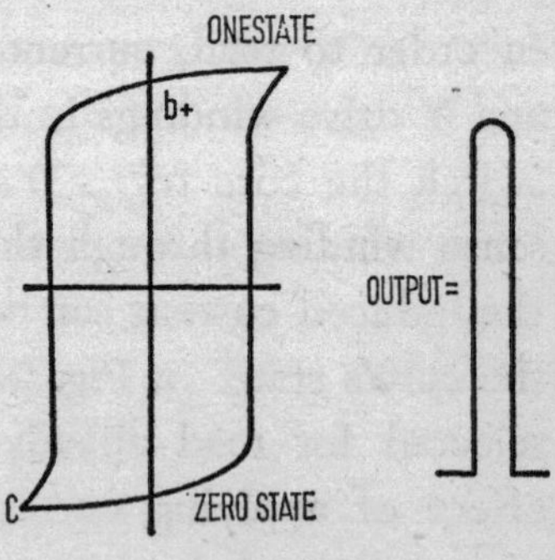

FIG 2.7

Because of the small physical sizes obtainable (less than 0·1 in optimum diameter) the cores are used extensively in the large main memories of computers, where it is necessary to have this storage available with capacities in excess of 128 000 (128K) 24-bit words.They are therefore assembled in a matrix of planes. A small single-plane matrix is shown in Fig. 2.8.

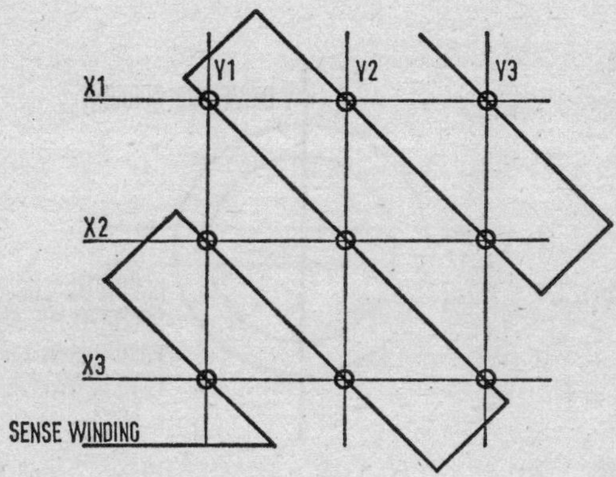

FIG 2.8

The X and Y lines are selected using some addressing system which differs from system to system. However, the method of writing requires just a little further explanation before we leave the philosophy. The write action to any selected number of bits is achieved by applying write currents to all X and Y drives in such a direction as to write 1's and in those positions where it is required to have 0; then a half-drive current is applied to yet another wire threading the cores and the effect will be to cancel out one of the half-write currents, and therefore the core will only experience a half-write current and will not switch (Fig. 2.9).

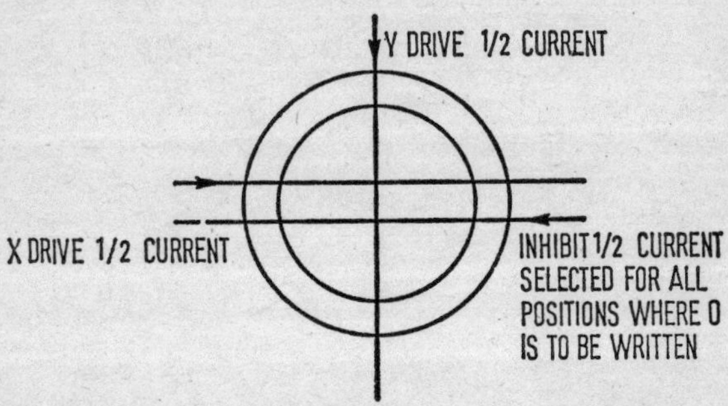

FIG 2.9

Other magnetic storage devices include (1) magnetic tape, where the binary pattern is recorded as a series of small magnetised areas along the length of the tape and can therefore, when read, be discriminated by recognising polarity or frequency, and (2) magnetic discs and drums, again using the philosophy as for magnetic tape. There are others, such as delay line storage and electrostatic storage, but these are not so common as the first three, which are termed 'backing storage'. An example of a use for the two types of storage is if we imagine a computer which has to carry out a payroll function of Rate x Hours; then the instructions would be held in the core storage (or Immediate Access Storage), as they are common to every transaction, whilst all the employee details would be held on some form of backing store and read in as required. It is also usual that, once processed, the result is again placed on the backing store for future calculations on salary to date.

3 Logical and Electrical Design

These two terms are used extensively in computer design and a short definition is given for each. *Logical design* is constructed using established mathematical equations for all computer operations, whilst *Electrical design* translates these logical equations into practical and optimised electronic circuits.

The objectives in our design are:

1. To 'crunch' numbers for what is at best a numeric mincing machine. We therefore require a numbers system.
2. To describe mathematically all the operations to be carried out.
3. To represent all instructions and quantitative information by electrical quantities.
4. Develop electrical circuits to enable all this to be possible, including primary power supplies.

The first of these has been dealt with already by the section on Binary Arithmetic. The second is a development of Boolean Algebra, named after George Boole who introduced it around 1850. Although it does not give directly a method of designing the most efficient circuit, it does allow logical functions to be written as a mathematical expression, which may then be simplified.

Only two states exist in the algebra of logic: yes or no, true or untrue, present or not present–these are represented by '1' or '0' respectively. In electronic logic these states refer to different voltage levels. To be amenable to algebraic manipulation, the signals applied to a logic element must be essentially simultaneous, since time is not defined as a parameter of the algebra.

The third objective–'to represent by electrical quantities'–is achieved in differing ways; but as, for example, it is only required to represent two values, ON/OFF etc., an electronic pulse or level of potential is the most suitable means of representing these values. Therefore all information, whether control or data, is represented in the computer in the form of pulses or levels. This, then, is the internal language of our computer and we will define our pulse terminology.

In Fig. 3.1, we see a representation of two pulses with appropriate definitions:

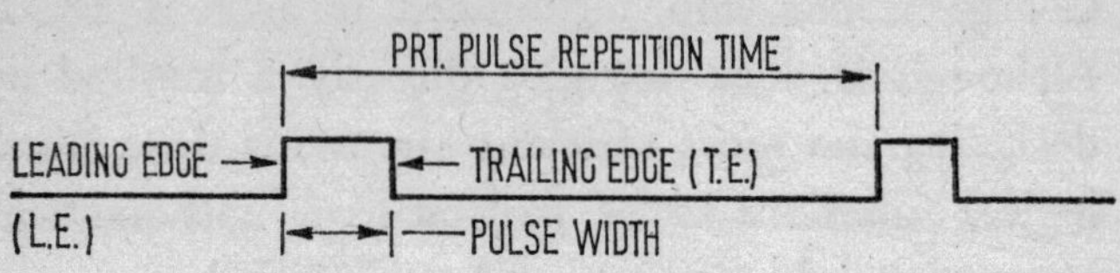

FIG 3.1

Of course, diagrammatic representation of pulses always tends to be idealised thus:

whilst the actual pulse often looks like this:

and therefore circuits in the computer, such as

amplifiers, pulse shapers, etc., are used to correct these waveshapes for accuracy in use.

It is necessary at some stage of the computer's use that synchronisation between various actions is achieved, and this is normally controlled by an oscillator, which is really an electronic clock producing regular pulses which can then be used to synchronise two or more actions.

Levels are more readily understood, but it is usually a pulse which changes the level as shown in Fig. 3.4.

FIG 3.4

Finally, the necessary circuits to put life into our machine are too numerous to discuss in a book of this type, as they revolve around technologies which, in themselves, have been the topic of many volumes. However, one golden rule is that the potentials and currents used are small, i.e. 0 V-3·5 V and currents in the order of milliamps; and it must follow that designers go to great lengths to produce stabilised, noise-free power supplies.

Boolean Algebra

As in ordinary algebra, we shall refer to numbers as 'x', 'y', 'z', etc. In this algebra, however, we must remember that such numbers can only take one of two values, i.e. 0 or 1. We must also think of a symbol, such as x, y, z, as representing a logical proposition which can be either true or false. We shall define the addition sign as indicating the OR logical function. For example, if we make the logical statement that proposition z is true if either proposition x or proposition y is true, we can write this in our symbolic algebra:

$$z = x + y$$

Since x and y can only take the values 0 and 1, it is quite easy to form a table showing the corresponding values of $z = x + y$ in all possible cases, and the table is obviously:

x	y	$z = x + y$
0	0	0
1	0	1
0	1	1
1	1	1

We shall also agree to define the logical AND function. In this we state that proposition z is true provided that both propositions x and y are true. We make the function of writing this symbolically as: $z = x \times y$. It is quite easy to make a corresponding table for the product xy as follows:

x	y	$z = xy$
0	0	0
1	0	0
0	1	0
1	1	1

We define the quantity 'not x' as x', so that if $x = 1$, $x' = 0$, and vice versa. It will be obvious that

$$x \times x' = 0$$

and that $$x + x' = 1$$

The above two statements are two of the additional algebraic rules we referred to earlier. As far as manipulation of symbols goes, all the ordinary rules apply as well. It is possible to prove any rule in Boolean algebra by constructing a table on the lines above, e.g. one of D'Morgan's rules:

$$(xy)' = x' + y'$$

A further logical function which is useful in computer circuitry is

$$z = xy' + x'y$$

This is the logical expression of the fact that x and y are not equivalent, so that z represents 1 if x and y are different, and 0 if they are the same. It is written symbolically:

$$x = y$$

We illustrate by the following table:

x	y	x'	y'	$xy' + x'y$
0	0	1	1	0
1	0	0	1	1
0	1	1	0	1
1	1	0	0	0

We must now translate these logical notions in terms of electronics. The electronic device referred to as a 'gate' has already been mentioned in Chapter 1, and this must now be elaborated further. The two basic types of gate are AND gates and OR gates. Logical designers use the symbol in Fig. 3.5 to denote an AND gate.

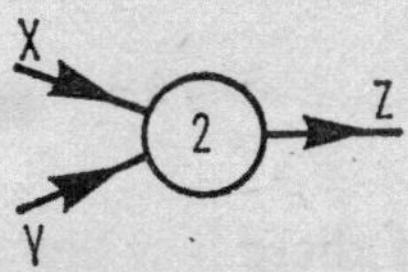

FIG 3.5

In this system a signal appears on the output line z if, and only if, signals are present on both the input lines x and y; this is indicated by the number 2 written in the symbol. In terms of Boolean algebra, we can write $z = xy$. This type of gate can be generalised by having more than two input signals, and in such cases is symbolised as shown in Fig. 3.6 below, where $z = x \times y \times w$. Gates of any number of inputs can be constructed.

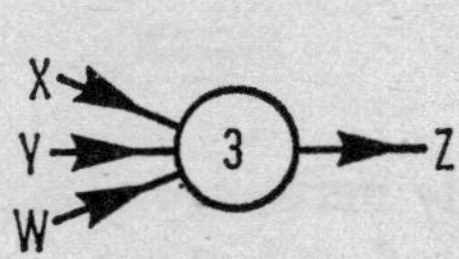

FIG 3.6

An OR gate, denoted symbolically below (Fig. 3.7), is constructed whereby a signal appears on

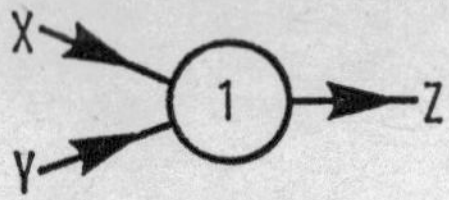

FIG 3.7

the output line if there is a signal on either of the input lines. This is indicated by the number 1 in the symbol, and the algebraic expression for the operation of this gate is $z = x + y$. In the same way as for AND gates, we can, of course, have more than two inputs. A three-input OR gate as shown below (Fig. 3.8) represents the algebraic expression $z = x + y + w$.

FIG 3.8

It is frequently necessary to have a negating circuit which transforms the number x into its negative x or not-x.

The truth table for this is clearly

x	x'
0	1
1	0

Symbolically we represent this

Here the single input x emerges as the single output x'.

As an example, suppose we wish to generate the output $x + y'$ from inputs x and y.

This can be achieved simply as follows:

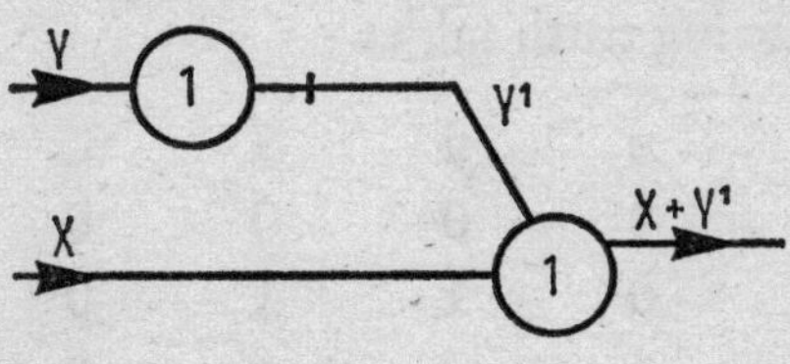

FIG 3.10

Addition

Let us consider the logical problem involved in the design of a binary adder.

Let us add 7 = 0111 to 5 = 0101:

7	0111
5	0101
12	1100

The details of this operation, starting from the right are:

1 + 1	= 0 with 1 to carry
1 + 0 + 1 (carry)	= 0 with 1 to carry
1 + 1 + 1 (carry)	= 1 with 1 to carry
0 + 0 + 1 (carry)	= 1

And so we arrive at the correct answer of 1100 or 12 in decimal.

If we consider the result of adding two digits only in all possible combinations we can arrive at the following truth table:

x	y	S	C
0	0	0	0
0	1	1	0
1	0	1	0
1	1	0	1

It must be stressed that this table applies to the addition of two digits only. If there are more than two, as above, we frequently have three binary digits to add, since we must include any carry from the previous place.

If we study the above truth table we notice that the carry C is simply the product $x \times y$, while the sum S is the function $xy' + yx'$. A logical diagram which will generate these two functions C and S is shown in Fig. 3.11.

Note that the logical scheme generates the function $S = (xy)'(x + y)$.

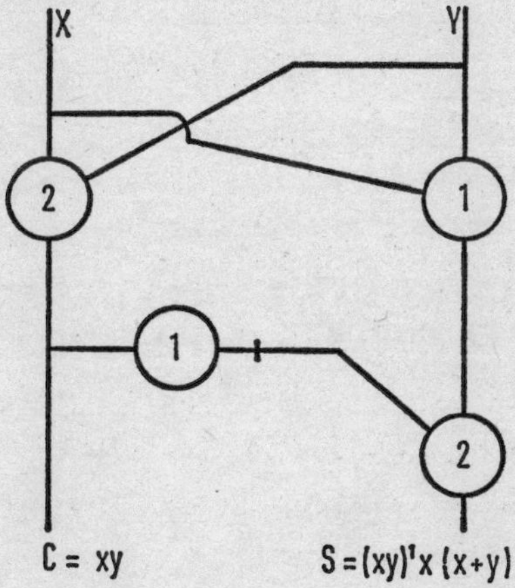

FIG 3.11

A simple transformation in Boolean algebra shows that this is the same thing as $xy' + x'y$.

$$\begin{aligned}\text{For}\quad (x+y)(xy)' &= (x+y)(x'+y')\\ &= xy' + yx' + xx' + yy'\\ &= xy' + x'y\end{aligned}$$

since $xx' = yy' = 0$.

The logical circuit in Fig. 3.11 is called a Half-adder and is denoted frequently by the symbol:

FIG. 3.12

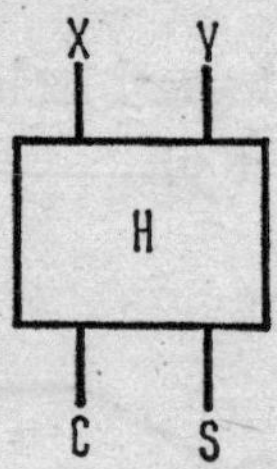

This is because it only does part of the job. A Full Adder must take account of the carry from the previous place.

A truth table for a full adder would be

x_r	y_r	C_{r-1}	S_r	C_r
0	0	0	0	0
1	0	0	1	0
0	1	0	1	0
0	0	1	1	0
1	1	0	0	1
1	0	1	0	1
0	1	1	0	1
1	1	1	1	1

We can generate this function by combining two half-adders as shown in Figs. 3.13 and 3.14, p. 58.

Representation of Negative Numbers

In order to fix ideas, let us take a register containing six binary digits. We can represent sixty-four different numbers in the register from 000000 = 0 to 111111 = 63. In ordinary arithmetic we indicate negative numbers by placing a minus-sign symbol in front of the number, the operation of subtraction being carried out by using a set of rules. This system would be highly inconvenient in a computer. Fortunately, there is another way of representing negative numbers, which is to replace them by their *complement*. We can define the complement of a number x with respect to a radix N as $N - x$, thus the complement of 7 in respect to 10 is $10 - 7 = 3$. The usefulness of this concept lies in the fact that we can subtract by adding the complement, e.g. if we wish to subtract 7 from 9 we add the complement of 7, i.e. 3, getting:

9

3

12

57

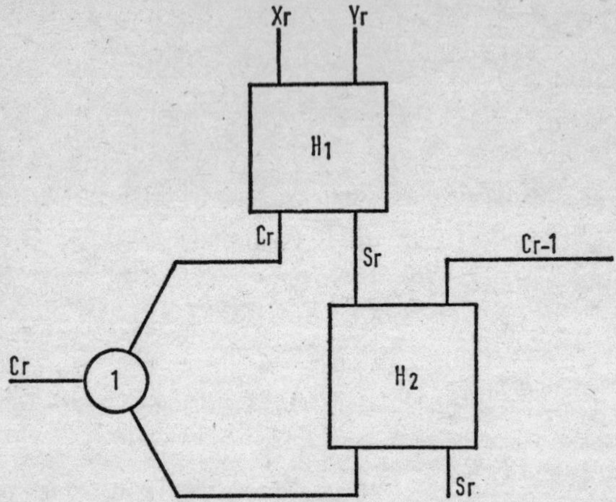

FIG 3.13

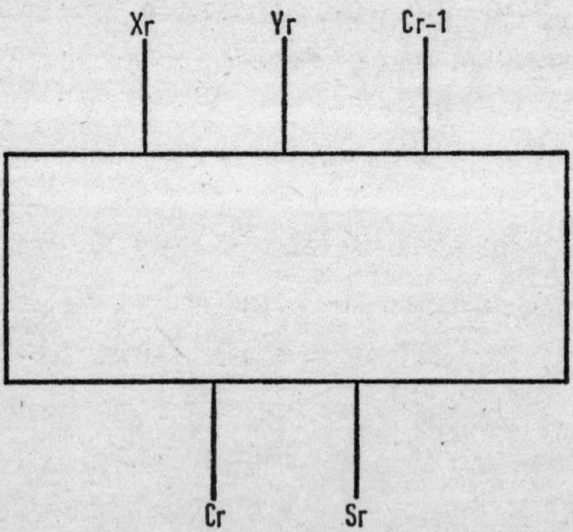

FIG 3.14

The answer appears to be 12, but we must regard the 1 as having 'spilled over' from the register, which, in this case, is thought of as having one decimal digit only. Had we been using a three-figure register, we would have had to take the complement of 7 with respect to 1000, and the operation would proceed:

$$\begin{array}{r} 009 \\ 993 \\ \hline 1002 \end{array}$$

Here the answer is as before, and once again the 1 has 'spilled over' to the left side of the register. Now suppose we wish to subtract 8 from 3; the complement of 8 is 2 and the answer is:

$$3+2=5$$

Here we get the correct answer 5, but there is nothing to indicate that it is *minus* 5. Fortunately, the binary system once again comes to the rescue. Taking a six-figure register, the complement of any number x is 2^6-x, i.e. $64-x$, and must be expressed in binary form. Let us subtract 19 from 26. The complement of 19 is 45 and the arithmetic is thus:

$$
\begin{array}{rr}
26 = & 011010 \\
45 = & 101101 \\
\hline
 & 1000111
\end{array}
$$

The answer is 7 in binary with a 'spill over' as before.

Now let us try to subtract 26 from 19. The complement of 26 with respect to 64 is 38. We now get:

$$
\begin{array}{rr}
19 = & 010011 \\
38 = & 100110 \\
\hline
57 = & 111001
\end{array}
$$

Not only do we appear to have obtained the answer 57, which is wrong, but here too there is no indication that the answer is negative. The correct answer is, of course, −7. We note, however, that if we take the 64 complement of the above answer (57) we get the number 7, but there is still nothing to show that it should be −7.

Now let us agree that if the first digit of a binary number is 0 the number is positive, and that if the first digit is 1 the number is negative and its numerical value is its 64 complement. This reduces our positive numbers in a six-figure register to lie between 0 and 31, i.e. 000000 and 011111.

Negative numbers must now lie between 100000 and 111111. Thus at the cost of restricting a six-figure register to hold numbers lying between −32 and +31 we have solved the sign difficulty. In the example above our answer was 111001. This by definiton is negative, and it is the 64 complement of 7. Hence it represents −7, which is correct. Some further examples will make the point clearer.

(1) 001011

This is positive because the first digit is a zero and its numerical value is 11.

(2) 100111

This is negative because the first digit is a 1. Its value as it stands is 39 but, since it is known to be negative, we must take its complement with respect to 64 to get its actual value in our convention. This is 64 – 39 or 25. So the number above is –25.

(3) 101111

This is negative. The face value is 47. Its complement with regard to 64 is 17. Hence it is –17.

(4) 011011

This is positive and its value is 27.

If now we wish to reverse the process and express decimal numbers in the convention there is no difficulty with positive numbers. We merely express them in binary form, making sure that the first digit is a 0. (This restricts us to 31 as a maximum in a six-figure register, as already explained.)

e.g. $$19 = 010011$$

Before we can do anything with negative numbers, however, it is first necessary to find their complement (in our case the 64 complement). There is an easy method for doing this which we shall now describe.

Formation of Complements in Binary Mode

Take any number in the binary system, e.g. 011011, and invert it, getting 100100. Clearly if we add these together we are left with 111111, so it is correct as far as a six-figure register is concerned to write:

$$x + x' = 2^6 - 1$$
$$2^6 - x = x' + 1$$

i.e. to form the complement of a number x invert it and add 1. Not only is this a simple operation arithmetically but it is also a comparatively easy matter to invert a number by electronic means, and therefore the construction of an automatic complementor which electronically inverts a number and adds 1 to it is easy to achieve.

Example. Subtract 13 from 28.

$$28 = 011100$$
$$13 = 001101$$

But we need the 64 complement of 13. We find it as follows:

	13	=	001101
Invert this and get			110010
Add 1			1
64 complement of 13		=	110011
Now add this to 28			011100
			110011
			1/001111

The answer is + 15 if we ignore the spill-over.

Example. Subtract 23 from 5.

$$5 = 000101$$
$$23 = 010111$$

Find the 64 complement of 23.

23	=	010111
Invert this and get		101000
Add 1		1
64 complement of 23	=	101001
Now add this to 5		000101
		101001
		101110

This is negative, and its numerical value is 46.
Take the 64 complement, which is 18.
So the answer is −18, as it should be.

Multiplication

Multiplication consists in the pure binary mode of a series of adding and shift functions.

Let us multiply 19 by 13.

```
   010011            19
   001101            13
   ------            --
   010011            57
  000000            19
 010011             --
010011              247
---------
011110111
= 247
```

The shifting of numbers is a simple electronic operation, and we have already explained how to deal with addition. We are, of course, restricted in the size of numbers in order to ensure that the answer does not exceed the capacity of the register. This method of shifting and adding applies only to pure binary systems; for other systems a method known as halving and doubling is used. The principle is that we write down the two numbers and proceed successively to halve one and double the other. In the halving column we will in general rapidly have an odd number. In this case we reduce it by one before halving again.

Let us carry out this process for 13 and 19, the numbers discussed above.

	Halving column	*Doubling column*
(Subtract 1 before halving)	13	19
	6	38
(Subtract 1 before halving)	3	76
	1	152

When we get to 1 in the halving column we stop.

If it were not for the odd numbers which keep coming up in the halving column the correct answer would be the figure in the doubling column opposite to 1 in the halving column. This is because the product of two numbers

$$x \times y$$

is identical to $\frac{1}{2}x \times 2y$

or $\frac{1}{4}x \times 4y$

etc.

and when we end with a 1 in the left-hand column the product must be the figure in the right-hand column. We must note, however, that, when there is an odd number in the left column, by subtracting 1 before we halve again we have thrown away the number opposite the odd number. We must, therefore, add it back. Hence we should get the correct answer by adding the right-hand column numbers opposite odd numbers in the left-hand column to the final figure opposite the 1.

	13	19
	6	(38) ignore this
	3	76
	1	152
We get		247

which is the correct answer.

Since 1 is itself an odd number, we can make a rule of operation, which is to carry out the halving and doubling process and then add up all numbers on the right opposite odd numbers on the left, ignoring the numbers on the right opposite even numbers on the left.

Try another example.

56	x	15	ignore
28		30	ignore
14		60	ignore
7		120	retain
3		240	retain
1		480	retain
		840	

which is correct.

This method of multiplication is said to have been employed by Russian peasants, and is really an example of binary multiplication, even though it is applied in the decimal system. If we take a close look at the binary multiplication process above we see that the answer is obtained by adding the following:

010011	=	19
0000000	=	0
01001100	=	76
010011000	=	152
011110111		247

So that what looks like a straightforward decimal-type multiplication in binary is, in fact, a direct case of halving and doubling.

In machines which use the binary coded decimal system, electronic means of halving and doubling have to be devised, and form the basis of the multiplication system. In this case the numbers to be multiplied, which may be held in two registers, are passed decimal digit by decimal digit to a halver and a doubler respectively. We can arrange for the number in the doubler to be passed through an adder if the number in the halver is odd, and the total accumulated in a third register.

Floating Point Arithmetic

Quite clearly, in any machine we are limited in the size of the numbers we can handle by the size of the registers. For example, the normal capacity of the registers in a commercial computer is twelve decimal digits. It may happen that, when a series of calculations is being performed, we have a need to use larger numbers at some stage, even accepting that the final answer can only have the number of significant figures corresponding to the capacity of the register. Another instance may be in, so to speak, wasting register capacity in representing a small decimal number with several zeros before the significant figures. A system by which these difficulties can be surmounted is described as the floating point system. In this we define a number by giving two sets of digits. In the decimal system this works as follows:

7 345 000 = $\cdot 7345 \times 10^7$ and would be specified by the pair 7, ·7345.

0·000 067 25 = $\cdot 6725 \times 10^{-4}$ and would be shown as −4, ·6725.

Thus we could handle both these numbers in a five-figure decimal register if we adopted the floating point convention. Clearly, special steps have to be taken by the programmer to allow for the system. The same principle can, of course, be applied to numbers expressed in binary form, but in this case the 'decimal' point means that we continue in descending powers of 2, not of 10.

Engineering Design

This is far too complex a subject to be handled in depth by a book of this sort, but some obvious constraints are observed. As computer design has advanced the hardware of machines has become more miniaturised and standardised, with the result that the logic designer is now able to place more of his hardware on one 'package', which consists of a printed circuit board where the various power rails are produced by layers of conductive material with insulating layers in between. But the world of computers is a practical one, and therefore the logic designer has to be guided by the design engineer, who must take into account the economics involved in fault finding on the end-product as well as a multitude of other problems. It is possible to mount all the components of a powerful computer on one board and, as time is money with the majority of computer users, then the system of trouble shooting which has evolved in most instances is one of replacement and repair at leisure. Therefore there must be an economic limit to the size of packages until development has perfected all components.

One important tool used by computer designers today is, naturally enough, their own product, the computer. He can use software (programs) to recognise his problems and optimise the results of his design.

The system which has been evolved to assist in this way comes under the title of Design Automation, and other by-products from the software involved are such things as printouts of the logic diagrams, wiring information and component listing.

Another related problem is that, due to the degree of miniaturisation possible, component siting and wiring can give rise to interaction, which has to be carefully considered. Where electrical noise and induced effects cannot be practically reduced by re-siting or re-routing, then non-logical elements have to be incorporated to correct this interference, e.g. shapers and amplifiers.

Control Sequences

Let us for a moment look at a simple data processing action—that of a cashier checking out goods at our local supermarket. She is following a set of mental instructions—pick up goods, check price on item, ring up value on cash register, repeat action until all items rung up, then press total key, present bill to customer, etc. A computer has to carry out these sorts of actions, and many other varied and complex sequences, and it is therefore necessary that the instructions are stored somewhere in the machine, and acted upon in sequence. This is achieved by storing, in the memory, unique patterns to represent our instructions, e.g.

1 0 1	0 0 0 0 0 0 1	00001010000001
A	ADD	B

which can instruct our machine to ADD contents of address B to those contained in A, leaving the result in A.

Of course, 100 or more of these unique instruction patterns will be used on an individual computer, so we must utilise in our machine some form of instruction or function decoder so that the necessary control lines and routing paths are active for the instruction being handled. This can be achieved by having an instruction/function register, where, during the instruction selection phase, the instruction is deposited and the outputs from this are decoded, to access the addresses holding the operands and then define the particular operation to be performed.

4 The Control System

This is probably the hardest part of a computer to understand, and it tends to be glossed over in lectures and elementary descriptions. We shall try to show how a computer control system could work and not to describe an actual system which does work. The latter would lead us into too much detail and would make the principles which lie behind all control systems much harder to grasp. To fix ideas, we shall describe a very rudimentary computer which would only be capable of carrying out very simple operations, but the principles in its control will apply to more complicated systems. Let us assume that our computer has two registers, P and Q, and a magnetic drum store. The only arithmetical operations which it can perform are addition and subtraction. At this stage the means by which information can be got into the com-

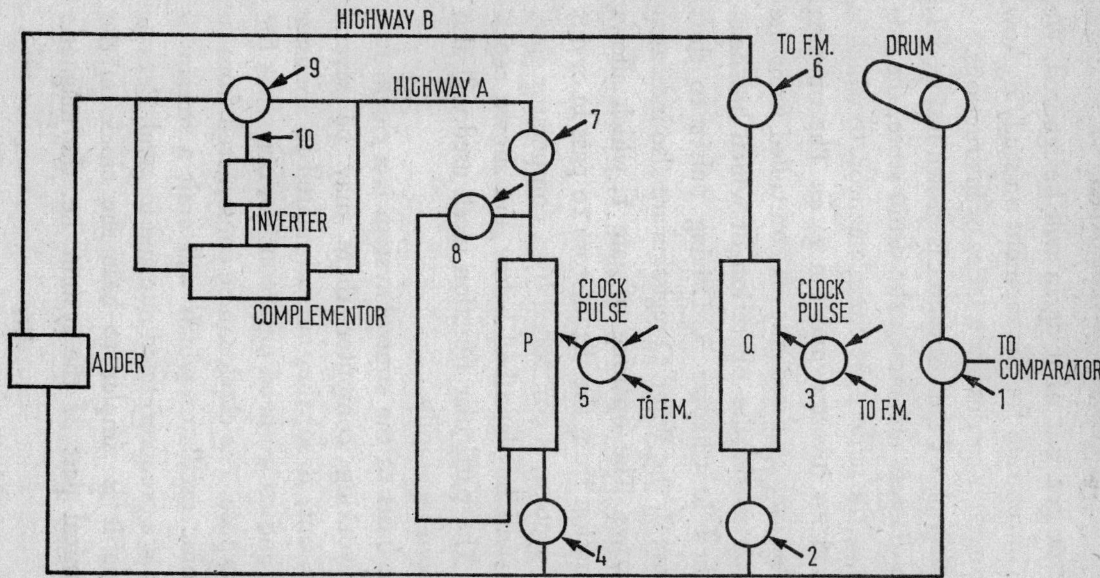

NOTE – ○ ↖ TO FUNCTION MESH

FIG 4.1

puter or got out of it again will be ignored. We shall concentrate for the moment on studying how the computer carries out a series of instructions.

A diagram of the computer is shown in Fig. 4.1. In the diagram are shown the drum store, the two registers, the adder, the complementor, to which is attached an inverter, and ten gates. The opening and shutting of these gates is controlled from the function mesh, the operation of which has been described in Chapter 3. Driving pulses to shift numbers in the P and Q registers are also indicated by arrows. The opening of gate 1, which allows information from the drum store to pass in or out of the system, is controlled by a comparator gate, which will be described below and shown separately. The particular function mesh used will also be shown separately.

The basis of the control system is a clock. In a real, working computer there may be several clocks, but, if so, they are all related to a master clock, and so we need only consider one clock for our purpose. The clock can be constructed from an electronic oscillator which will emit a stream of pulses at a predetermined rate and interval. For our purpose it is simpler to base the clock on the mechanical part of the system, i.e. the magnetic drum. If we cut notches on the edge of the drum these can be made to generate electronic pulses as the drum rotates. This has the special advantage that it automatically synchronizes everything else to the drum's speed. Even if the drum speed varies slightly, no errors can occur, as the clock timing is adjusted accordingly. Let us assume that we have eight tracks on the drum and sixteen words, each made up of sixteen bits, on each track. We require a space between words, and the space will be that required to hold four bits of information. Thus, as the drum revolves we will get a succession of pulses, sixteen at a time, separated by a space corresponding to the time for four pulses. We require one special pulse to indicate the beginning of each revolution. The spaces are required in order to provide a little time for opening and shutting gates, and to allow other parts of the system to settle down before the next word is ready to be dealt with.

Let us concentrate for the moment on a particular track. As the drum revolves, a reading head reads each word in turn as it passes under the reading magnet. This word cannot, however, escape into the computer system unless a gate is opened to allow it to do so. In order to ensure that the correct word is allowed out of the store, the programmer first decides which word he wishes. For example, let us suppose he wishes the word number 'thirteen'; the number 13 is set up in binary form in a flip-flop register, alongside which is a counting register which counts the words as they pass. This counter is, of course, driven by the clock. Just before the word 'thirteen' comes up under the reading head, the counter shows the number 13 in binary form. We now have the two registers in coincidence; this causes a gate to open and word 'thirteen' is allowed to flow out into the computer as it is read. The next word could be number 'fourteen'; there is no longer any coincidence, and the gates are shut. It is easy to see that by this means we can select any word we wish from the drum; all that is necessary is to wait the necessary time until the correct word comes under the reading head. The control flip-flop, as it is called, contains, of course, more information than the number of the word on a particular track (coded in binary form) together with the number which is the identifying code for the function which we wish to be carried out. We shall assume in our rudimentary computer that we have ten functions numbered 1 to 10 as follows:

1. Transfer word from address *a* on drum to P register.
2. Transfer word from address *b* on drum to Q register.
3. Add the number in P register to the number in Q register and put the answer in Q.
4. Clear add. (This means that we allow the number in P register to pass through the adder and drive the number in the Q register out, leaving Q clear to accept the number from P. This, in effect, transfers from P to Q.)
5. Subtract P from Q, putting answer in Q.
6. Clear subtract (i.e. put complement of P in Q).
7. Test Q for positive or negative value. If positive take next instruction from address *x* on drum, if negative take next instruction from address *y* on drum.
8. Transfer from register Q to address *c* on drum.
9. Test Q register for zero. If answer is 'yes' place word in location R in Q register, if 'no' transfer word from location S to Q register.
10. Stop.

These ten instructions are sufficient to enable us to write a simple program.

In order to perform any calculation by computer it is necessary to store somewhere the necessary constants involved in the problem. For example, in a set of equations we need to have in the store the numerical coefficients of the unknown variables. It is also necessary, of course, to store the program instructions so that these can be transferred as requested to the control register and subsequently obeyed. We shall assume that this has been done in the case of our rudimentary computer, and that the program instructions and any constants required have been placed in known locations in the drum store. Since we have only two registers, we shall have to do frequent transfers to and from the store, even when we wish to store a number only temporarily. Since we have not assumed any form of input or output, we must signify success or failure in our calculation by transferring some pre-arranged pattern or number from the drum to one of the registers. These patterns must also be put in the store.

It is helpful to illustrate the problem by doing a simple mathematical calculation on our elementary computer. Let us try to solve the problem of finding whether a given number is or is not a perfect square. If it is we shall give its square root. It is necessary to proceed in successive steps as follows:

Mathematical Investigation

Quite apart from any question of using a computer, we must first see whether this problem can be tackled mathematically. There may be more than one method, but we must choose the most straightforward one in order to give the computer the most elementary arithmetical problems to deal with. One obvious way is to subtract in turn all numbers which are perfect squares. If we eventually end up with a zero answer, then our number is a perfect square. If we end with a negative answer, it is not. This method would require a knowledge of squares, and by looking up a table we might find the number we are studying, which makes the problem a trivial one. We must assume that no such table exists. Let us take the numbers 64 and 57 and try out a method which we could apply in the computer. Instead of subtracting 1, 4, 9, 16, etc., in succession, it amounts to the same thing if we first subtract 1 and subsequently, instead of subtracting 4 from the original number, let us subtract 3 from the first remainder. We then take 5 from the second remainder and so on. Note that we increase the number to be subtracted by 2 each time. This number 2 must be one of the constants which we hold in our store. The arithmetic is shown below for the two cases of 64 and 57.

Note that if every time we achieve a successful subtraction, i.e. a positive or zero remainder, we score 1, then the sum of the accumulated 1's is the square root in the first case and the nearest but next least integer in the second case.

On page 83 we show how to draw what is called a flow chart to illustrate the various steps in the calculation. It should be noted that each step in the flow chart may represent several program steps. It will be seen from the flow chart that the program repeats itself so long as the remainder is positive. If, however, we achieve a negative remainder, we adopt an alternative branch in the flow chart and, therefore, in the program in which we add back the last number and test whether the answer is zero or not. If it is zero we transfer a prescribed number from the drum to Q register and then stop. The appearance of this number indicates that the number is a perfect square. If the answer is not zero we transfer another number from the drum to P register and then stop, this number indicates that we do not have a perfect square. In

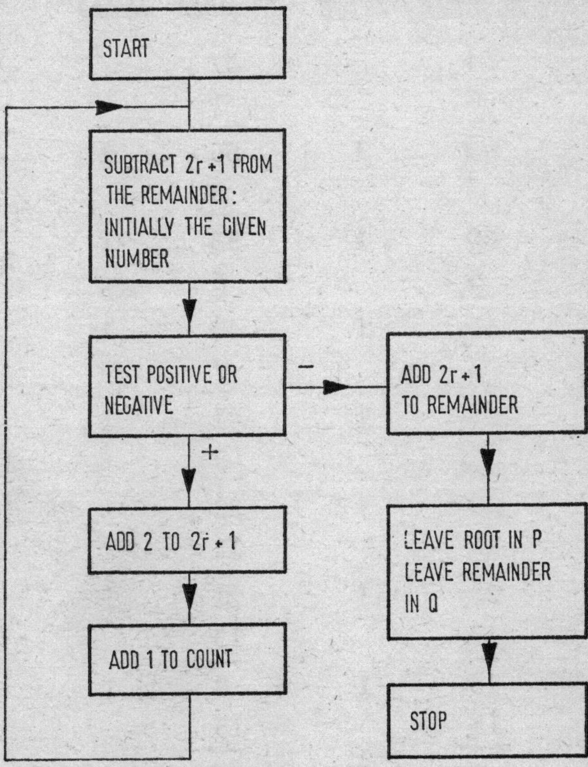

FIG 4.2

practice, of course, a computer at this stage would print the answer. As we have not included any output in our simplified computer, we must indicate the result in the registers.

Number	*Score*	*Number*	*Score*
64		57	
1		1	
63	1	56	1
3		3	
60	1	53	1
5		5	
55	1	48	1
7		7	
48	1	41	1
9		9	
39	1	32	1
11		11	
28	1	21	1
13		13	
15	1	8	1
15		15	
0	1	−7	
	8		7

The detailed program steps for this operation are given below in order to illustrate the detailed nature of program steps. Note that every eventuality has to be forseen and nothing must be left out. In a real computer, with many more functions and with input and output equipment, a program may run into dozens of pages.

Program

A, B and C are drum locations which store the odd number, remainder and count respectively. Initially A is assumed to hold the constant 1, B the number which is being inspected, and C the constant zero. The program is then:

	Step	*Instruction*	*Remarks*
	1	Transfer B to Q	Forms next remainder
	2	Transfer A to P	
	3	Subtract	
	4	Test positive or negative	
If positive	5	Transfer Q to B	
	6	Transfer A to Q	Forms the next higher positive integer
	7	Place constant of 2 in P	
	8	Add	
	9	Transfer Q to A	

	10	Transfer C to Q	Add 1 to count
	11	Place constant of 1 in P	
	12	Add	
	13	Transfer Q to C	
	14	Jump to step 1	
If	15	Add	Put square root in P and the remainder in Q
negative	16	Transfer C to P	
	17	Stop	

If Q is zero, the initial number was a perfect square and its root is in P. If Q is non-zero, the remainder is in Q and the root of the next lower square is in P.

The Control Register

The Control Register holds the program instruction which is to be obeyed. A numerical code is used, and this must be expressed in binary form. The Control Register must contain:

1. The function to be performed, e.g. add, subtract, shift, etc.
2. The source, i.e. the location of the number on which we wish to operate.
3. The destination, i.e. the location in which we wish to put the result of carrying out the function in 1.
4. The address of the next instruction.

As we have already seen, our rudimentary computer has ten functions which require four binary digits. The address of the number on which we wish to operate requires seven digits, and the same applies to the destination. The address of the next instruction will also contain the same number of digits. Thus the number to be stored in the control register contains eighteen digits as follows:

*	f	f	f	f	t	t	t	w	w	w	w	a	a	a	a	a	a	a

* Four bits, indicated by f f f f, define the function.
Three bits, indicated by t t t, define the drum track.
Four bits, indicated by w w w w, define the word.
The full address of the next instruction on the drum obviously requires seven bits, indicated by a a a a a a a.

When an operation is complete a signal is sent and the next instruction is read into the control register. The machine then carries on with this instruction. In some cases we can simplify the process by having the address of instructions in serial order. In this case we need only add one to the address of the last instruction to get the next one. The basic function of the control register is to open gates, which then admit driving pulses to cause the necessary function to be performed.

5 Storage Devices

The chapters preceding this have dealt with the physical characteristics of the hub of computer systems. This is that part of the computer in which all the calculation work is carried out and is known as the Central Processor, a term which is fairly self-explanatory. The following chapters deal with a computer system in toto.

A computer system is that which, by computerised methods, can perform four distinct functions:

1. Input of data.
2. Processing of that data.
3. Storage of results.
4. Output of results.

To date we have studied the processing function, thus it now remains to look at input, storage and output.

The three functions are achieved by 'plugging in'

(interfacing) to the central processor a series of peripheral devices which will perform those functions while remaining under the control of the central processor. Obviously any device which stores information for the central processor must have the means of transferring data to and from it. It could therefore be categorised as an input/output device. In order to arrive at a suitable division of peripherals, the author has assumed that any device forming an interface between the human element and the computer is input/output, and therefore dealt with under Chapter 6. The remaining devices which essentially can only impart their information to humans via a secondary process are therefore storage devices and dealt with as such below.

Reading and Writing

Before studying storage devices it is necessary to clarify their operation. Storage devices have an overriding requirement for speed. Since the speed of processing information is equal to the speed of an electric impulse completing one 'lap' of the processor, then storage devices should be equal to this, or at least not very much slower. To this end, therefore, information is not punched into, or printed on, storage devices but is recorded on them using electromagnetic techniques. Thus there is no translation phase involved and the whole process is entirely computer-oriented. The process of putting information into store is known as '*writing*'. Writing information will destroy any information which exists without the necessity to erase first. This can easily be pictured in the case of the factory lights (Chapter 2). In order to change from Smith to Jones it is not necessary to switch off all the lamps first: it is merely necessary to apply the code for Jones. Thus lamps already on will be unaffected if they are to stay on, and only those not required will be switched off. The same, in reverse, applies to lamps switched off for Smith.

The process of extracting information from store is known as *reading*. Reading information does not destroy the original, in much the same way that reading this book does not destroy the printed pages. There is a slight difference, however, because a human has 'memory'; if the same information is required on a subsequent occasion, the human can recall it from memory. Since the computer does not have this facility, it becomes more important that reading does not destroy data, since for any subsequent use of the data it will have to be re-read.

Immediate Access Stores

These are ultra high-speed stores which are contained within the central processor. Such a store is extremely costly and therefore kept to a minimum. The cycle time (time taken for one impulse to complete one circuit of the store) is often measured in nanoseconds (ten-millionths of a second). Immediate Access Stores are of two types.

Magnetic-core Stores

The property magnetic hysteresis, discussed in Chapter 3, is seen in operation in this form of storage. In order to fully understand the operation of Core Store we need to relate hysteresis to the application. The situation can be represented in Fig. 5.1 in which the magnetising force H is plotted horizontally, and the resulting state of magnetisation in the substance, called B, vertically. B is normally much larger than H because of the alignment of molecules inside the material, which turn themselves into elementary magnets, thereby contributing themselves to the magnetic flux inside the

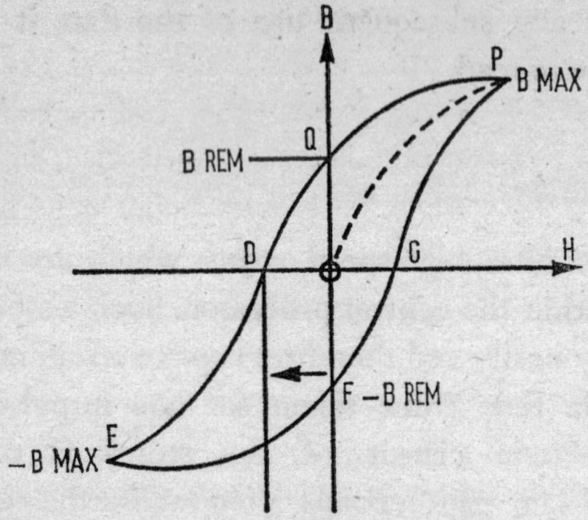

FIG 5.1

material. If we start with our material in a magnetically neutral state represented by the point O on the diagram, and then increase the applied magnetic field H, we see that B at first increases, but after a time it reaches a maximum value B_{max} represented by the point P, and further increase in H has no effect on B. At this stage the material is said to be saturated. The first sign of hysteresis occurs when we start to reduce the applied field H with a view to removing it. Our substance does not retrace its early path on the diagram but follows the path PQ. This means that when the applied field has been removed the material remains in a magnetised state represented by the point Q. The value of B at this point is B_{rem} (for remanent). If we still further reduce the applied field by making it negative we can take the substance to the point D at which B is zero. Further decrease of H takes us to the point E, where we again reach saturation in the opposite direction, and $B = -B_{max}$. If we now start to increase H once more we arrive at the point F corresponding to a value of $B = -B_{rem}$. Further increase in H takes us to the point G when $B = 0$, and we eventually arrive again at the point P when $B = B_{max}$. Note that we do not pass through the origin at O. We have now traced out a closed loop, and by retracing our variation in the applied field H, we retrace our loop. This loop is referred to as the Hysteresis loop.

The usefulness of this property depends on the fact that certain substances have a Hysteresis loop which is very nearly a rectangle. This means that, if we apply a positive field of sufficient strength and then remove it, we shall leave our substance in the state of magnetisation represented by the point Q. If we then apply a negative magnetic field of sufficient strength and remove it, the substance will settle at the point F. We have here two states of magnetisation, and these can be used to represent a 0 or a 1 in the binary system. The state of magnetisation of the substance is stable and will persist indefinitely until removed by an externally applied magnetic field. Furthermore, the transfer from one state to the other can be achieved at very high speed by applying a short but intense pulse of magnetising force. When a piece of magnetic material changes its state it is said to be switched. A switching time of 1 microsecond is common. No substance has an ideally rectangular loop, but some are sufficiently near rectangular to be used in the above manner.

If we cast such a substance in the form of a hollow ring or core, and thread this with a conducting wire, the applied magnetic field can be produced by a pulse of current through the conducting wire. Reversal of current will reverse the magnetic state of the core. Fairly large currents are necessary to produce the requisite magnetic field, and in some cases several turns of fine wire are wound round the core. This reduces the current required, but it is an expensive operation from the production point of view and necessitates a core large enough for the threading to take place. Because of these drawbacks, wound cores are no longer in common use.

The real advantage of a core as a storage device dates from the discovery of the principle of half-current writing. Suppose that we thread two single wires through a core, and suppose that the current required to switch the cores is I, what happens to a core if we pass a current $\frac{I}{2}$ through one wire? If we assume that the core is in the residual state represented by point F in Fig. 5.2, a half-current pulse will change the state of the core to the point represented by J, but on removal of the current pulse the state will revert to that corresponding to the point F. If, however, we pass two half-currents through the core, then it will be carried to a point near to P, and on removal of the current pulses the cores will now return to the point Q. Consider next Fig.5.3 on page 97.

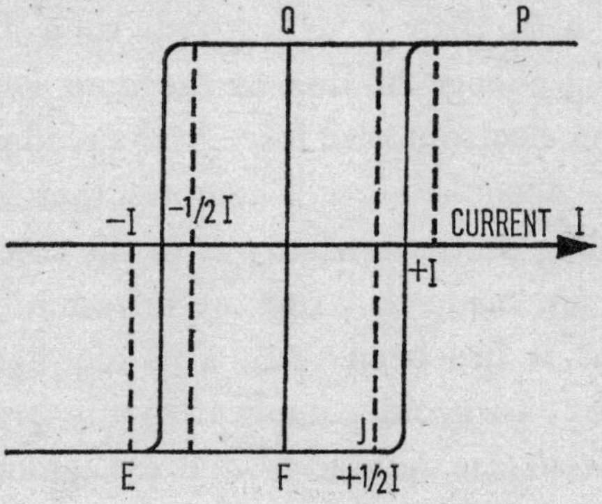

FIG 5.2

If we now pass half the current required to change the state of a core through all the horizontal wires in row X and the same half-current through the vertical wire in column A, then the only core which will receive a full current will be the one at the intersection of row X and column A. This core will 'switch' and the others will return to their former state, since they have only had a half-current. This gives us a means of writing a 0 or a 1 in any core we wish or, in other words, we can send information to an address of our choice.

If we wish to read what is in the core we need a third wire which may be threaded through all cores. Suppose we apply a pulse so as to write a 1 in our core. If it already holds a 1, nothing will happen and our read wire will show no response. If it has a 0, then it will switch to a 1, and the resulting change of flux as the core switches will cause an electromotive force in the read wire which we can recognise. It will be noted that this process of reading destroys what was in the core. Arrangements are made to write the information back as soon as it has been read. This requires a fourth wire, but to avoid complicating the principles of the system it is omitted from the diagram.

Due to the fact that no substance has a perfect

loop, there is, in fact, a small response induced in the reading wire from the other cores in the appropriate rows and columns. There are ways of competing with this, one of which is to thread the reading wire so that these spurious signals cancel out as far as possible. In a very large store other means may have to be used.

Cores used in this manner are very small, since there is no winding problem.

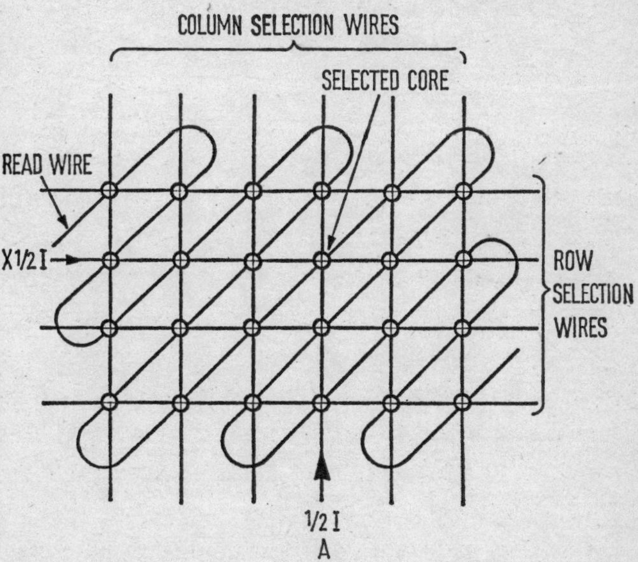

FIG 5.3

Thin-film Stores

This is a very recent innovation for use in cases where the amount of information required within the central processor is low but where a phenomenally high processing speed is required. The technique is to replace the ferrite core with a plastic base coated with a fine film of magnetic oxide. Due to the greatly reduced resistance of this type of store, speeds can be increased to fractions of a nanosecond, but, because of the exceptionally fine tolerances in manufacture, the cost of such a store is sufficient to prohibit its use in general commercial computers.

Buffer Stores

Buffer stores are essentially a device for gaining time or for reconciling the different time scales inherent in a computing system. For example, the speed of feeding and reading the information on a punched card is limited by mechanical factors. Even with the faster types of card reader it is difficult to exceed 1500 digits per second. In the absence of any storage device, we have to do any calculation in the time available between reading one card and starting to read the next one. One way of gaining time is first to transfer the information on a card into a buffer. This enables the computer or calculator to process the first card while the second card is being read by the input device. We have thus gained a whole card cycle for a calculation. There are innumerable applications of this principle. We normally wish to read tape at the full speed of the machine. This is far too fast for a printer. So we read the tape into a buffer store and print from the buffer.

In older, or second-generation, computers, buffer stores were a separately identifiable section of the immediate access store, often with a different cycle time. In some instances buffers would be contained on a high-speed drum and would not really be integral to the processor store at all. However, with increased computing speeds inherent in third-generation machines and slow peripherals still limited by mechanics, the requirement for buffering has become more pronounced and has lead to buffering becoming a programming technique of allocating areas of immediate access store rather than having physically separate buffers.

Magnetic Storage Devices

We have seen how we can make use of the property

of magnetic hysteresis to record information on a length of suitable wire.

The other principle which we shall require in order to understand magnetic methods of storage is Faraday's principle of magnetic induction. In simple terms this states that, if a conductor comes under the influence of a varying magnetic field, an electromotive force is induced in the conductor. For example, if we had a closed loop of wire and we brought a bar magnet nearer to the wire, more lines of magnetic force would thread the loop. This would induce an electromotive force in the loop, which would result in a flow of electric current which we could detect by a sensitive measuring instrument.

If we now suppose that we have a second electromagnet, when a portion of the wire which has been magnetised passes under this magnet it causes a change in the magnetic flux through the coils. This change of flux induces a current in the coil, and this can be detected. It is in this sense of detecting that a portion of the wire has been magnetised that we describe a second magnet as 'reading the information'.

The first magnet which magnetises the wire is said to 'write' information on it. These similar magnets are usually referred to as 'reading' and 'writing' magnetic heads.

By making the air gap smaller, say 0·001 inch, and keeping the heads about the same distance from the wire, it is possible to store information at a density of about 500 bits per inch. The faster the wire moves past the heads, the greater the signal received.

This type of storage is clearly limited in its usefulness by the fact that one could have to search through the whole length of wire to retrieve one piece of information. However, in spite of this fact, a serial access backing store is still probably the most widely used method, although wire has now been replaced by tape in order to achieve higher speeds and a greater packing density (up to 1600 bits per inch).

Magnetic Tape

Magnetic tape makes use of the same principles as magnetic wire. Instead of iron wire, it is the practice to use a magnetic oxide which is sprayed on to a plastic film of tape. The thickness of the film is about 0·001 inch. Because of its width, which is normally 0·5 inch, it is possible to put information on several parallel tracks, as they are called. Each track requires its own reading and writing heads. Tape does not run continuously but as required by the computer. It is mounted on what is called a Tape Deck. In this device a reel of tape can be transferred to a second reel by an electric motor. In going from one reel to the next, the tape is made to pass under reading and/or writing heads. The direction of motion can be reversed. We cannot begin to read or write on the tape until it has reached full speed. Because of this, information on tape has to be written in blocks, with a space between each block to allow for starting and stopping. Various mechanical means exist for achieving this rapidly. The usual practice is to accelerate the tape from a loop only and to use the length of the loop to actuate a servo motor, which then unwinds more tape to keep the loop approximately constant in length. The length of the loop can be measured by some kind of photoelectric device. Large accelerations are necessary, a common type of specification being from 1 inch per second up to 100 inches per second in 2 or 3 milliseconds. This limit applies to stopping as well as starting.

Applications of Magnetic Tape

Reels of magnetic tape can be quite long–2400

feet is not uncommon. A reel of this type with eight parallel tracks can store 14 million numerical or alphabetical characters. It follows that, in order to gain access to a particular piece of information, it may be necessary to run through most of the reel, and this can take as long as a few minutes.

It is thus customary to use tape for the storage of information which can be updated serially. Examples of this are Insurance Company records or Customer Name and Address files.

Magnetic tape has the secondary advantage of being much cheaper than any other form of backing storage, both in itself and in the cost of tape decks. Therefore it is common to find magnetic tape being used to hold duplicate copies, for security purposes, of files held on Direct Access Devices.

Direct Access Storage Devices

The method of recording on these devices is exactly the same as on magnetic tape. However, the method of accessing that data differs significantly. It is as well to look at the access method generally before going on to describe the individual devices.

During the section covering magnetic tape we

discovered that the data was written and read serially; this has the inherent disadvantage that one could be forced to read through the whole of a reel of tape in order to access one item of data contained on the end of a reel. Fairly quickly it was agreed that this constituted a vast amount of wasted computing time and that a method was required of holding data on one continuous flat surface so that any item of information could be accessed directly.

The early direct access stores were a coated cylindrical drum which was divided into up to sixty 'tracks' horizontally. A read/write head was positioned over each track and the drum revolved beneath them. Thus, by activating the appropriate read/write head at the correct time, one could directly access any piece of information contained thereon. As the forerunner of present devices the drum was eminently successful, and such was its speed that a drum was used for buffer storage in many second-generation computers.

However, it was quite costly and rather restricted in that every track of data required its own read/write head.

Disc storage followed quite quickly. Here data is recorded on a disc resembling a gramophone record, and each disc surface has one read/write head.

To access data the disc is firstly divided up into concentric circles, each of which forms a 'track'. The read/write head can be positioned over a track to pick up data. However, in order that one specific item of information may be selected, rather than having to read through a whole track, the disc is divided down even further. The tracks each contain a number of 'blocks' and data is recorded into these.

In order to read or write directly, one first allocates to each data record a unique key by which it may be referenced. The program which uses this data holds an index, which contains the record key against the block number into which it has been written. The central processer control system holds a secondary index, which shows block numbers against the track which contains them.

Thus we have the basis of current direct access devices.

Magnetic Drum

Although one of the first direct access systems, the drum is still in operation, but restrictions of physical size, cost and capacity have seen it largely superseded by other direct access devices.

Magnetic drum has a capacity of up to 2·5 million characters and a transfer rate of around 1·4 million characters per second. Thus, although exceptionally fast, its capacity by comparison with magnetic tape (10-14 million characters) is low.

Fixed Disc Store

This is a method of holding large quantities of data in a direct access environment.

Up to thirty discs, each approximately 3 feet in diameter, are held on a horizontal spindle. A set of multiple read/write heads is positioned between each disc surface. The driving motor for the spindle is generally held centrally along it, with the discs arranged either side of the motor (Fig. 5.4).

Arrangement of data on the discs is in accordance with the general description given earlier in this chapter.

The time taken to access a given item of information on this type of storage device is subject to three factors:

1. Seek Time: the time taken for the read/write heads to traverse from one track to another.

2. Latency: the time required for the revolution of the disc to bring the data record beneath the read write heads.
3. Verification Time: this is a security check built into the disc mechanism whereby after any head movement a series of addresses is read and compared to ensure the head has come properly to rest.

Since all the above are governed by the physical speed of the machine, maximum, minimum and average times may be calculated for them and due allowance made in program suites.

Fixed disc stores have capacity to hold up to 450 million characters and can transfer these at speeds approaching 150 000 characters per second.

Fixed Disc Store is so named because the individual discs are not interchangeable.

Exchangeable Disc Store

This is the 'little brother' of fixed disc. It differs in that this type of disc is mounted horizontally on a vertical spindle. Furthermore, in order to arrive at

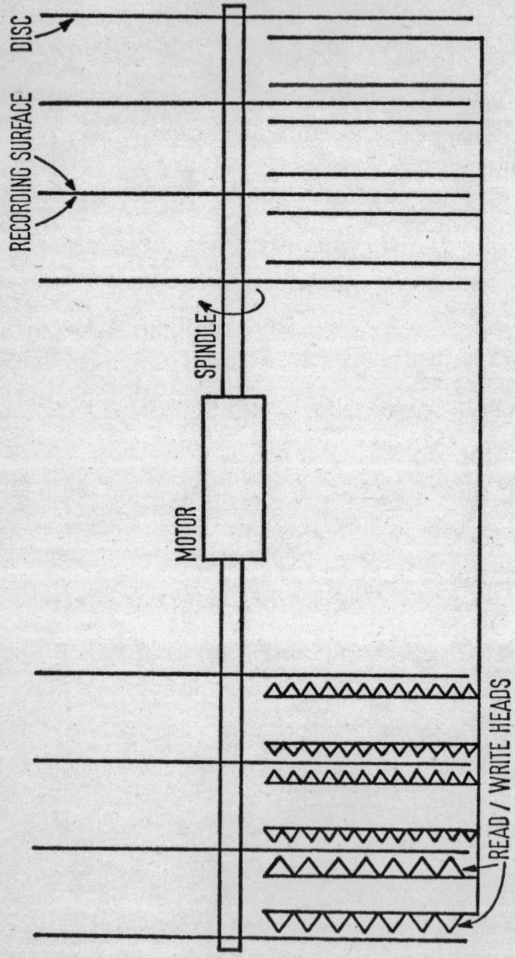

FIG 5.4

interchangeability of discs, the recording surface has been reduced to 14 inches diameter and the number of discs to between two and twelve, depending on capacity specification. Exchangeable discs are supplied in cartridge form, and it is a simple manual operation to change cartridges on the disc transport.

To facilitate exchange, single read/write heads are positioned between the discs (Fig. 5.5) and these retract into the transport casing as soon as the motor is de-activated.

Recording methods and timing considerations

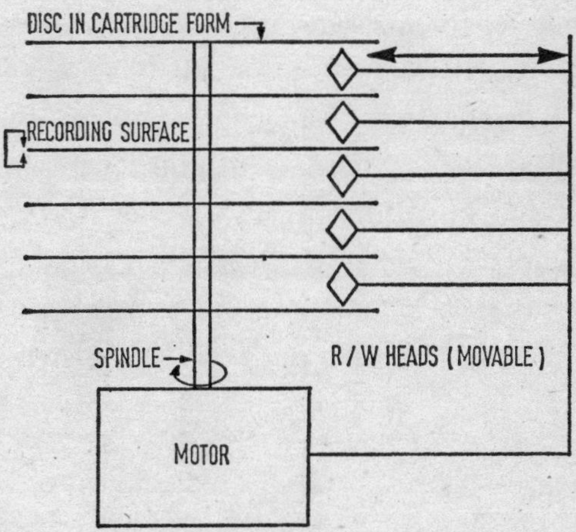

FIG 5.5

are as already described, and the capacity of these devices varies from 0·5 million characters (two-disc) to 60 million characters (twelve-disc) per cartridge. The rate of transfer of data to and from exchangeable disc stores is of the order of 210 000 characters per second.

Direct access devices operate through a high-speed controller, which is in essence a series of buffers between the transport and the central processor. By using such a device it is possible to link several transports to one data channel, thereby considerably enhancing the direct access capacity of a given configuration. However, when using one data channel in this way, it is not possible for data to flow both ways simultaneously, which in real time situations (see Chapter 9) may be a serious disadvantage. Therefore, most configurations with a multiplicity of direct access transports would contain at least two controllers, each connected via a different data channel to the central processor, thus allowing simultaneous reading and writing of data.

Applications of Direct Access Devices

There are two major criteria to be considered in deciding when to use direct access as backing store:

1. The urgency with which data is required.
   In a communications system the user is demanding instant response to his enquiries. Since magnetic tape may keep the enquirer waiting for several minutes before responding, direct access devices are regarded as mandatory.
2. The 'hit rate' of file processing.
   When updating a file (e.g. adding recruits and deleting leavers from a personnel file) only a certain percentage of records are updated. By comparing the average number of changed records with the total number of records on file, one can arrive at a 'hit' percentage where each updated record is classified as a hit. If the resulting percentage is low, then significant time savings will accrue by using direct access in place of magnetic tape.

6 Input and Output Devices

In this chapter we shall study the various devices by which humans can put information into the computer and receive information from it.

Input Devices

These are the means by which the basic facts of a problem are fed into a computer. We shall mention punched cards, paper tape, magnetic tape, optical character readers and the console typewriter.

Punched Cards

A typical punched card is shown in Fig. 6.1.

It will be noted that there are eighty columns, and each column has twelve possible positions or rows in which a hole can be punched. Ten of these positions are for representing numbers from 0 to 9, and there are two additional positions, frequently

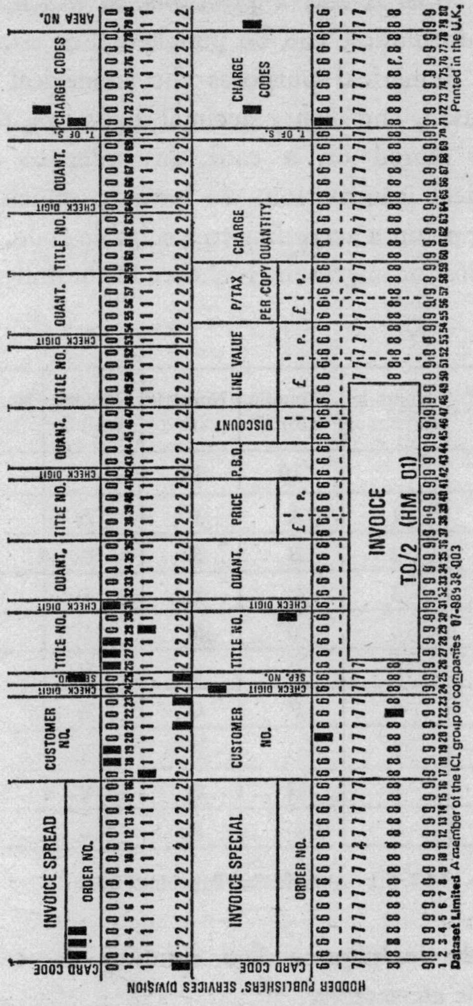
INVOICE SPREAD
ORDER NO.
CUSTOMER NO.
TITLE NO.
QUANT.
CHECK DIGIT
CHARGE CODES
AREA NO.
INVOICE SPECIAL
PRICE
LINE VALUE
DISCOUNT
P/TAX PER COPY
CHARGE QUANTITY
INVOICE
TO/2 (HM 01)
HODDER PUBLISHERS' SERVICES DIVISION
Dataset Limited A member of the ICL group of companies
Printed in the U.K.

FIG 6.1

referred to as X and Y positions, in which additional information can be punched. For straightforward numerical purposes the numerical part only is used, and eighty decimal digits can, therefore, be stored on a card. In order to store alphabetical information, we have to punch two holes per column according to a suitable code.

A typical arrangement is shown in the following table:

Numerical Punching only	Numeric (no zone)	Numeric with X	Numeric with Y	Numeric with O	Numeric with I
		10	11		
1	1	A	J	&	
2	2	B	K	S	%
3	3	C	L	T	¼
4	4	D	M	U	–
5	5	E	N	V	/
6	6	F	O	W	½
7	7	G	P	X	
8	8	H	Q	Y	÷
9	9	I	R	Z	¾

FIG. 6.1 Alphabetical Punching Code

This code includes a few symbols as well as alphabetic characters.

Cards are available in a variety of sizes. Small cards containing only forty columns are in common use, and ninety-six column cards are also employed. The eighty-column cards are probably the most popular. The hole punches in the card can be either rectangular, as in the diagram, or round. One advantage of the round hole is that it is possible to double up on the number of holes punched by punching additional holes between the rows. This makes the cards contain, in effect, 160 columns of information. The columns on the card are usually divided up as convenient into small groups known as fields, numbers in a field embodying a special significance, such as £p, etc., and the appropriate heading can be printed on the card. Some cards have several headings printed on them, such as 'Income Tax Paid to Date', 'Gross Pay', etc., so that the card can be read as a document.

Punching and Verifying Cards

The cards have to be punched by hand from information on original documents. This can be done with a simple mechanically operated punch with twelve keys. Two keys have to be depressed simultaneously for alphabetical characters. It is more usual nowadays to use a typewriter keyboard which is electronically connected to the punch knives. It is usual to verify the punching of a card, since cards, once punched, may be kept for several years and used over and over again. Accuracy is therefore essential. Verification is carried out by, in effect, repunching the card. Any discrepancy between the keys operated by the verifier and the characters punched in the original card causes the machine to give warning, and the card can then be checked by a supervisor.

Reading the Card

We mentioned in Chapter 1 that cards may be read by detecting the presence of a hole either by an electrical contact or by means of a photoelectric cell. Clearly, there must be a means of feeding the card to the reading mechanism, and cards are, in fact, picked out from a stack one at a time and pass along a track to the reading stations. There may be one, two or even three reading stations in order, for example, to compare one card with the next or to read certain control information before dealing with the whole card. Cards can be read either eighty columns at a time, i.e. in parallel, or twelve rows at a time, i.e. in series.

It will be obvious that the second method is the

one most suited to a computer which processes information serially. The speed of serial feeding is around 300 cards per minute, or roughly 400 characters per second. This is slow for a modern computer.

Parallel feeding can be faster, and speeds of 1200 cards a minute are not uncommon. In this case, however, the information has to be placed in a store and taken out in serial form as required.

Cards can also be punched in binary notation, but normally they contain information in decimal form, and this needs to be decoded into binary before it can be processed by the computer.

Interpreting the Card

It is very convenient to be able to read what has been punched in the card in plain language. This can be done by passing a punched card through a machine called an interpreter; this prints across the top of the card in alphabetical or numerical form what is punched in any column.

Paper Tape

Comparatively speaking, this is a fast means of presenting information, speeds of up to a thousand characters per second currently being achieved.

The tape is prepared by punching on a special hand punch or by the output of an enhanced electric typewriter. The tape is normally 1 inch wide and is punched with five, six or seven, but normally eight, holes across its width. An example of a paper tape code is given in Fig. 6.3 on page 120.

The tape is read by passing it between a light source and a series of photoelectric cells, spaced across its width, one per track.

Verification of paper tape still takes place but generally on the same machine as is used to prepare the input. This can be switched to 'verify' and the process is as follows.

The operator loads the punched tape into a reader on the machine and the blank tape into the punch. When a key is depressed, one frame of punched tape is read into the machine's buffer and compared with the character which has been keyed. If the two characters are equal, then that character is punched into the blank tape and the operator may proceed. If, however, the two characters are dissimilar, then no output character is punched, the keyboard locks and an error warning light glows. In this way the operator is made aware of the error condition. Across the top of the keyboard are two series of lights which, by illumination, show the codes of the character read and the key depressed (a miniature version of the factory lights—Chapter 2). The operator now checks the codes against the original. If she has pressed the wrong key, then she presses an 'error clear' key, which unlocks the keyboard and extinguishes the error light, and tries again. On re-depression no new character is read, but the new key depression is validated as above.

If, however, the character which has been read is incorrect, then the operator presses a 'correction' key, which causes the keyed character to be output and overrides the keyboard lock and error light.

Although sounding a long-winded process, all comparison is electronic and will not reduce the effective speed of the operator, except in error conditions.

Paper tape has acquired great popularity in recent years; mainly because of its low cost, and that of the equipment required to produce it. Also the physical inflexibility of the material has some advantages; for example, if young Florrie, the glamour-girl operator, is flustered by admiring looks from George, the Adonis of Accounts, while carrying 2000 cards and she drops them, then it is a major operation to re-assemble the cards in

	CARD PUNCH	PUNCHES 1	2	3	4	5
A	Y1	●	●			
B	Y2	●			●	●
C	Y3		●	●	●	
D	Y4	●			●	
E	Y5	●				
F	Y6	●		●	●	
G	Y7		●		●	●
H	Y8			●		●
I	Y9		●	●		
J	X1	●	●		●	
K	X2	●	●	●	●	
L	X3		●			●
M	X4			●	●	●
N	X5			●	●	
O	X6				●	●
P	X7		●	●		●
Q	X8	●	●	●		●
R	X9		●		●	
S	02	●		●		
T	03					●
U	04	●	●	●		
V	05		●	●	●	●
W	06	●	●			●
X	01	●		●	●	●
Y	08	●		●		●
Z	09	●				●
CARRIAGE RETURN					●	
LINE FEED			●			
LETTERS		●	●	●	●	●
FIGURES		●	●		●	●
SPACE				●		

	CARD PUNCH	PUNCHES 1	2	3	4	5	
1	1	●	●	●		●	
2	2	●	●			●	
3	3	●					
4	4		●		●		
5	5					●	
6	6	●		●		●	
7	7	●	●	●			
8	8		●	●			
9	9				●	●	
0	0		●	●		●	
&	01	●		●	●		
11	Y			●		●	
10	X		●		●	●	
+	X	●				●	PLUGGABLE FEATURE CREDIT + DEBIT
−		●	●				
3	Y	●					PLUGGABLE FEATUR 29−39 FEATURE
2	X	●	●			●	
1	0	●	●	●		●	

PLUGGABLE FEATURES

Y = 11d TWO COLS. IN TAPE

X = 10d " " " "

Y = 12 MONTHS TWO COLS. IN TAPE

X = 11 " " " " "

0 = 10 " " " " "

FIG 6.3

sequence and also to replace any which are damaged. This could cause an intolerable delay in the computer room. If, however, she had been carrying 2000 records on one reel of paper tape when George made his overtures, then it is but two minutes work to re-reel the tape and carry on.

Optical Character Recognition

This is a means of inputting source documents to the computer without having to go through the time-consuming and error-prone punching phases.

By shining light through a typewritten document a shadow can be cast. This is then passed through a complex optical system to enlarge it before arriving at a sensitised screen. This screen detects the outline of the character, which is then converted by control electronics into the correct code for entry in the computer buffer.

If the typed character is badly formed or so incomplete as to make it unrecognisable, then that part of the outline which has been received may be re-transmitted on to a visual display housed on the side of the machine. The operator can then identify the character and key it directly into the machine via a keyboard below the screen.

The drawbacks of this system at present are,

firstly, its very high cost; secondly, the restricted range of typefaces which it can recognise; and finally the fact that most machines can read only one line at a time at present. However, O.C.R. machines are developing very rapidly, and some of these factors may well be overcome before this manuscript is printed.

Console Typewriter

This is now mainly used as a means of enabling the operator or programmer to communicate with the computer. It has rendered unnecessary the former practice of having neon lights on the control panel to tell the operator the contents of the registers.

It is used during program testing and maintains a valuable record of all operations carried out. It can also be used to ask questions, such as the content of a particular store location.

Output Devices

Punched Cards

Originally devised as the output format of a calculator, card-punching devices are gaining popularity for their value as data collection media. For example, in a sales ledger situation a card can be punched with the account number and outstanding balance. The customer receives his statement together with the card and merely returns the card with his cheque. The card may then re-enter the system to adjust the account without further punching.

Paper Tape

This is gaining popularity as an output device for automatic machines. A computer program is used to analyse designs of machine tools and to break the design down into a series of operations to be performed by a paper-tape-driven machine. Each operation has a unique paper tape code which is punched, in the required sequence, by the computer. The resultant paper tape is then sent to the factory and read by the automatic machine to produce the parts required.

Printers

Printing devices are obviously the most popular form of output media, and several types of printers are available.

Line Printers. In order to speed up printing output, it is common to use machines which print a whole line at a time. This type of printer has been in use for many years in the classical tabulators. The earlier types were limited in the length of line printed and also restricted to numerical characters. The principle in this type of printer was to have a line of print slugs on a bar which could be raised vertically by a cam motion. The information read from the hole in a punched card produced an electrical signal which arrested the bar at the appropriate stage. After the card had completed its cycle, the print slugs were all stopped at various stages corresponding to the appropriate figure. A bar was then driven against the print slugs and the line was printed. By this means printing speeds of about eighty lines a minute can be achieved. In this sytem we are limited to printing numerical characters, and the next step was to achieve the printing of alphabetical information. This was done by putting the slugs on a rotating wheel so that all characters could be available by stopping the wheel at the appropriate time. If necessary, symbols in addition to alphabetical characters can be included. The selection of the character is done in two stages, by first arresting the wheel at a point corresponding to the appropriate numerical punching and then moving it in order to correspond to the second hole in the card corresponding to the alphabetical character. This means that the slugs on the print wheel must be arranged appropriately. The arrangement corresponding to the punching code in Fig. 6.2 is shown in Fig. 6.4. When the wheel has been stopped in the correct position, it is driven against the paper in the usual way to print

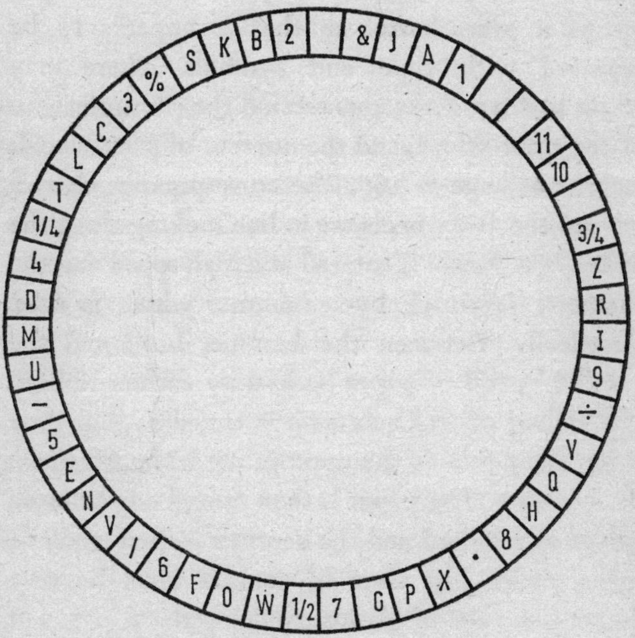

FIG 6.4

the character. Speeds of up to 150 lines a minute can be achieved, and the printing is of high quality, though without lower-case letters.

Hit-on-the-run Printers. The advent of computers stressed the need for greater and greater speeds, and one solution makes use of the so-called hit-on-the-run principle. In this the type wheels are keyed on to a print barrel so that it appears to be engraved with type and symbols. There may be up to fifty characters around the circumference of the print wheel, and the number of print wheels may be as large as 160. The corresponding characters on the 160 wheels are in line looking along the barrel. The barrel is rotated at a high speed and the character is struck by a hammer which is fired electrically. Between the hammer head and the barrel is a roll of paper backed by carbon ribbon. The falling of the hammers is timed so that they strike the paper as the appropriate letter flies past the hammer. The paper is then moved one or more spaces as required and the hammer is then ready to strike again. The time of contact must be very short if we are to avoid smudging the paper, and the timing of the hammers must be synchronised, or the print alignment will be poor. The principle of operation is shown in Fig. 6.5. We need rather more than one revolution to print a line, since one revolution of the barrel is required to print the necessary characters, and some additional time is required to move the paper to be ready for the next paper stop. Printers of this type are capable of speeds up to 1500 lines per minute.

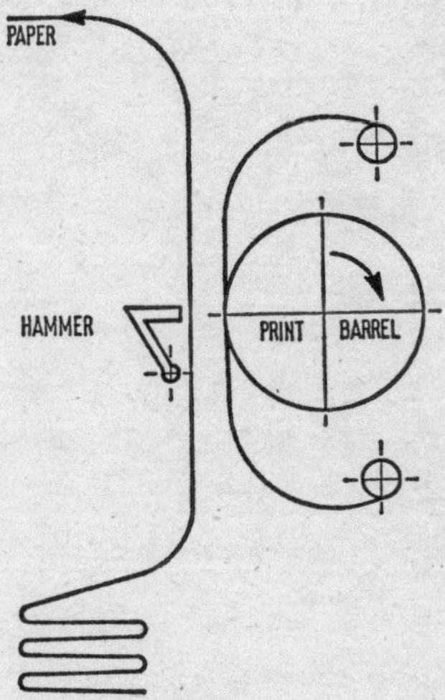

FIG 6.5

Xerographic Printer. The maximum speeds of mechanical printers are now being attained, yet printing still remains a comparatively slow process.

This means that, if printing is to be significantly speeded up, some other process is required.

One process which has had a lot of publicity recently is xerography. In this, dust of a suitable composition is attracted to the paper by an electrostatic charge. Instead of striking a hammer, we can pulse the hammer with a high voltage; this attracts the dust to the paper in the shape of the character required. By passing the paper under heated rollers, the characters can be fixed to the paper. Since there is no mechanical inertia to worry about, we can carry out the operation at very high speeds. Speeds of up to 80 000 characters per second have been achieved, but the system is subject to many snags.

Firstly, only one copy can be printed each time, although this has been partially solved by the development of in-line photocopiers through which the original can be passed immediately after printing.

Secondly, such equipment is extremely expensive, and so far only prototypes have been developed.

A practical problem which faces developers is paper movement. The paper has to travel at such a high speed through the printer that the only way of controlling it is to reel it at the output end. This means that the reel has to be unwound in order to commence reading!

Input/Output Devices

There are some devices which act as input and output devices. They have their own program in the computer which enables the operator to interrogate computer files and receive an immediate response. Because of this they are said to be 'on-line' to the computer.

Teletypewriter
This is the original on-line device, a printing keyboard not dissimilar to a typewriter in appearance and identical in concept. The teletypewriter is either linked direct to the computer (if local) or linked to a computer in another part of the country via telephone lines (if remote). The operator types an enquiry using the keyboard, and this also activates the hammers so that the enquiry is printed. However, the characters keyed enter a buffer within the teletype in much the same way as paper tape verifiers. When this buffer is full a signal is sent to the computer and this is registered by the control system, which sets a 'flag'.

The program which handles on-line devices has a 'poling' feature. That is, it will send out a signal to each device in turn when it has no enquiries to process. If the program identifies a flag for a device, it will then clear that device's buffer and process the enquiry. If no flag is set, it then poles the next device.

When the enquiry has been processed, the result is assembled in the central processor as a series of lines. These lines are then re-transmitted to the device whose flag was accepted. The impulses coming back to the teletypewriter activate the hammers, and thus the reply to the enquiry is typed out. When the message has been fully transmitted, the flag is unset and the next device in sequence is poled.

Here, then, is the ultimate usage of processor and direct access speeds. Using such a system it is possible to transmit replies to enquiries in fractions of a second, so the operator has 'immediate' response.

Visual Display Units

This is the most recent development in on-line equipment. Its method of operation is identical to the teletypewriter, except that the hammers, ribbon and paper are replaced by a small television screen. It is on to this screen that the characters are reproduced. However, if the characters were just placed once on the screen, they would rapidly fade and become illegible; thus, there is a device in the machine which re-transmits the buffer fifty to sixty times per second, thereby keeping the image bright and clear.

Such devices are quite expensive and require fairly large computers to operate. However, they can show vast savings in many directions.

For example, top management can use them to gather instant facts in order that truly up-to-date information can be used as a basis for decision-making.

They can be used as input devices and thereby remove the need for punching and verifying, since error checking can be on a visual basis.

The display terminal (visual display unit) is currently the 'in thing' with computers, and many applications such as booking agencies for theatres, holidays, aircraft seats, etc., are using them effectively.

7 Applications of Computers

Mathematical Applications

It must first be emphasised that computers cannot do mathematics or solve mathematical equations. They can only do arithmetic. There is a large field of mathematical problems for the solution of which so-called numerical methods have to be applied. In the past, desk calculators had to be used, and the speed at which these could be operated limited the usefulness of this approach, although in some cases a human computer might spend a year or more on the problem. The reduction of such problems to a set of arithmetical calculations is frequently a matter requiring considerable mathematical skill, and it can correctly be described as a branch of mathematics in its own right.

In many branches of science we are faced with problems for which present-day mathematical techniques are inadequate to provide an exact solution but for which a large quantity of observational data exists. As an example, let us consider the astronomical problem of the motion of three bodies under the gravitational forces each exerts on the other. It is easy enough to write down the differential equations whose solution will give the answer, but although this problem has received the attention of some of the world's greatest mathematicians in the last two centuries, no exact solution has been found. Happily for the users of the *Nautical Almanac,* it has been possible to find answers sufficiently accurate by approximate methods. For example, in calculating the motion of a planet we can regard the influence of the sun as predominant, solve as though it were a two-body problem and then treat the effect of a third body as a perturbation superimposed on the solution already found. Even this approach involves much laborious calculation, and when we realise that the solar system contains more than three bodies it is clear that ordinary desk calculators are out of the question. By making use of an electronic computer, however, and calculations based on tens of thousands of astronomical observations made over the last 200 years, Eckert and his colleagues in America were able to calculate the positions of five major planets—a six-body problem—at forty-day intervals up to the year 2000. This was done to an accuracy of fourteen decimal places.

In the realm of pure mathematics the computer is perhaps less important, but it can come to the assistance of the mathematician nonetheless. The construction of a table of prime numbers, which has engaged attention in the past, becomes extremely laborious as the numbers increase in size. By using a computer this table has been considerably extended. Another feat which is perhaps of interest only to the curious has been the calculation of π to 10 000 places of decimals. A further application to pure mathematics is on the solution of what are called Diophantine Equations. These are equations in which the solutions are restricted to whole numbers. Trial and error is one method of attack, and as the numbers tried become larger, recourse to a computer enables progress to be made.

A further class of problems which has recently received attention is that of the basically simple problem which involves so much arithmetic for its solution, which is exact, that it requires a computer. In this class we can include the solution of linear sets of equations.

It is sometimes possible to reduce a business problem to the solution of a set of ordinary linear equations of the type schoolboys learn to solve. The difference lies in the fact that we may have twenty or more equations to solve for the same number of unknowns. The arithmetic for this is basically simple but extremely laborious, and a computer can easily be programmed to do this calculation, provided that it has a large enough store to contain the initial data.

Linear Programming

This is a subject which has attracted much attention in recent times. The problem is to find solutions to a set of equations or inequalities, there being in general more unknowns than equations. This, of course, implies that there are an infinite number of solutions, but if we make the restriction that only positive solutions are admissible and then choose a set of solutions which makes a linear function a maximum or a minimum, a useful result can frequently be obtained. A simple illustration will make the point clear.

Let us suppose that we require to adopt a diet which contains minimum amounts of substances such as carbohydrates (C), fats (F) and protein (P). In order to achieve this we have the choice of several foodstuffs, such as fish, eggs, meat, cheese, etc. Let us assume that we buy x pounds of fish, y pounds of eggs, z pounds of meat and w pounds of cheese. By studying dietetic data, we can obtain the amount of carbohydrates, fats and proteins in one pound of each of these substances. Let c_1, c_2, c_3, c_4 be the quantity of carbohydrates in one pound of fish, eggs, meat, cheese respectively. Similarly, let f_1, f_2, f_3, f_4 and p_1, p_2, p_3, p_4 represent the fat and protein content. Furthermore, let us suppose that the cost per pound in pence is a for fish, β for eggs, γ for meat and δ for cheese. Then, if the permitted minimum of carbohydrates is C, of fats F and of protein P, we can lay down these conditions:

$$\begin{aligned} c_1x + c_2y + c_3z + c_4w &= C \\ f_1x + f_2y + f_3z + f_4w &= F \\ p_1x + p_2y + p_3z + p_4w &= P \end{aligned}$$

and in addition we wish to minimise the cost, viz.,

$$ax + \beta y + \gamma z + \delta w \text{ pence}$$

It is obvious that all x, y, z and w must be positive quantities, i.e. we need not buy any cheese (w is zero) but we cannot have a negative amount. This is a very simple illustration of a linear program problem. The solution can be obtained by trial and error using methods very similar to those employed in a set of linear equations. More practical problems of importance, however, involve a large number of variables, and solution by repetitive trial and error can be extremely laborious. It is not always possible to obtain a unique solution, but a number of successes have been secured, notably in the field of transportation.

The Monte Carlo Method

In certain cases mathematical problems may be solved numerically by setting up a model from which results can be obtained by observations obtained by random processes. In some respects there is a comparison between the process and certain games of chance, and it is for this reason that the method has acquired the designation Monte Carlo. The principle can be made much clearer by a simple illustration. If in a square of unit size we draw a curve PQ, as shown in Fig. 7.1, we can make an estimate of the area under the curve PQ shown hatched by the following process. If one were, for example, to throw a dart at the

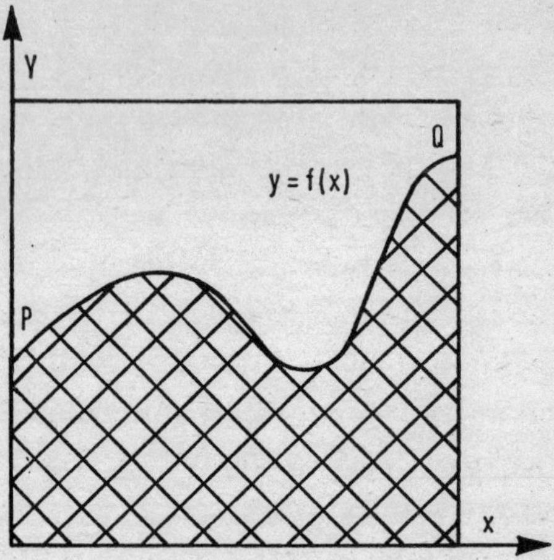

FIG 7.1

square, the probability that it would land in the hatched area is proportionate to the hatched area. Some estimate, therefore, of the size of the hatched area would be obtained if we threw a large number of darts, say 1000 or more, and counted those which fell in the hatched area. Suppose this number were 350, then a first estimate of the area would be 0·35 units, since the square is of unit size. This number is, of course, the ratio of 350 to 1000. Mathematically we could arrive at the same result as the actual experiment by selecting pairs of random numbers each lying between 0 and 1, and counting those whose position in the square was in the hatched area, which is defined by the curve PQ. In order to obtain reasonable accuracy, very large numbers of trials would be required, and in such a simple case alternative methods of finding the area would yield a much quicker result. If, however, we consider more complicated problems involving many variables, then the Monte Carlo approach is of some value. Until recently it had remained a mathematical curiosity because of the time involved in evaluating the answer, and also because, in order to increase the accuracy of the calculations, the labour increases out of all proportion.

A computer can handle calculations of this type in a reasonable time.

Engineering Applications

It frequently happens that design calculations for an engineering problem involve a large number of parameters which must be chosen in order to meet the specification laid down. The normal method of attack is to guess suitable parameters and then to perform a series of calculations to see if the specification is satisfied. If it is not, as is frequently the case, the parameters are varied and the calculation is repeated. Given an experienced designer, the correct parameters to give the specification may be found after one or two trials, but it may be that several trials are required. A computer is particularly suited to this type of repetitive calculation, since the program can be unaltered but the initial conditions varied and the calculation repeated. Having obtained the specification required, if a computer is available it is possible to proceed much deeper into the problem and to make further variations with a view to finding the minimum cost. This deeper application would generally be out of the question because of the time involved if the calculations had to be done by manual means. This type of calculation has been successfully applied to the design of large generators which are required to supply output to a known specification and also to the design of bridges which must carry a certain load.

Commercial Applications

The use of computers in commercial applications is now very much commonplace and almost at saturation point. One can read in the newspapers every day of examples of 'the computer' sending out an erroneous bill or 'the computer' sending a final demand for £0·0p and, of course, having to wait three days for a bank statement because of 'the computer'.

Commercial applications are now mainly regarded as being 'bread and butter' work, i.e. that which can be programmed and can start running fairly quickly to show early justification of the capital expenditure prior to commencement of more sophisticated techniques.

The reason for this attitude is that calculation involved in commercial operations is simple in comparison with mathematics and scientific work, the only requirement being to perform the basic calculations a very large number of times. Some examples are given below.

Gas and Electricity Billing

A common type of application is found in nationalised industries—for example, in the issue of gas and electricity bills. The calculation from meter readings of a quarterly gas bill is a very simple matter, even though it may involve different rates and instalments for equipment. One must bear in mind, however, that in some cases several hundred thousand such bills may have to be issued each quarter. The commercial computer is particularly suited for this type of work, since it can receive the information regarding meter readings on punched cards, or in more recent systems the meter reader can mark the card with an HB pencil and the marks are read and converted into figures by a machine such as the Optical Character Reader. These cards can then be fed into a computer programmed to calculate the amount due and to produce a statment of account for posting to the customer, often at a rate exceeding six thousand statements per hour. The program will then store the relevant portions of information in an accounts file. Using the data collected from previous quarters' statements, estimates can be produced for those customers who were 'out when the gasman called'. On the cash collection side, the statement contains a tear-off slip which the customer returns with his cheque. With optical character reading this slip can be read directly by the computer and the balance outstanding will be cleared. At the end of an appropriate period (e.g. monthly) a program can be run which will check the balances on the file. Any which remain uncleared will be the subject of a computer-produced reminder letter or Final Demand.

Payroll

One of the most common applications of computers to business is the production of a payroll. In this case the calculation is rather more complicated than in the electricity or gas billing problem because we have to take account of overtime, which is normally at a different rate from the ordinary hourly one; income tax has to be taken into account and subscriptions to benevolent or other clubs deducted. The preparation of payslips is an immensely complex operation and employs a large number of clerks. This, perhaps, is because monthly paid employees have their salaries paid into their own banks by credit transfer, and weekly paid employees may require cash and a payslip. The job can be done much more expeditiously from basic information by a computer, thus eliminating much tedious labour. A further advantage is in increased accuracy, which results in fewer pay queries. It is also possible to obtain information as to labour costs etc. very much more rapidly, and this information is generally required in budgetary forecasts, costing exercises and the like.

Stores Accounting

The computer has been applied to the problem of stores accounting by some large organisations, notable examples being the General Post Office, Regions of British Rail, oil companies, H. M. Armed Forces, etc. Such stores may contain over 100 000 items, each with its own reference number and with a total cash value running into millions of pounds. Stores may be scattered at various depots throughout the country. Every week there may be tens or hundreds of thousands of 'movements' in and out of the store. It is possible by means of a computer not only to record such movements and to have available at any time the value of stock held but the rapidity with which the information can be made available centrally also enables ordering and re-stocking to be undertaken in an economical manner. If one has sufficient confidence in the information available to reduce the stock level of a £10 000 000 store by £1 000 000, the interest on such a sum alone may pay for the computer installation.

Banking

This area has served to illustrate possibly more than any other the advantages of being able to transmit data from one part of the country to another for computer processing. Because of the nature of banking, a large high-speed computer is required with large amounts of direct access storage. Obviously the little country bank branches would not be able to afford the computing power they require. Therefore banks have installed very large machines at computer 'centres'. Each branch can then have a card reader and a printer installed and connected via telephone lines to 'the centre'. The branches are now able to transmit their data via cards, have it processed on the computer and the results re-transmitted to the bank printer. In this way even the smallest of banks can utilise computer control of accounts.

Production Control

If we consider a large production unit, the aim of the manager must be to produce his goods as quickly and as profitably as possible. The problem is so complex, however, that, even if a system based on human effort backed by great experience is successful, it is incapable of reacting other than very slowly to a changing situation.

It is such changes that are the main problem in running a production unit. Orders are cancelled or increased; labour troubles or material shortages may occur unexpectedly; design alterations arrive—it may be in hundreds; Company policy may alter. All or any of these factors tend to disrupt and delay the control system.

If one attempts to devise a system which can cover such contingencies, it is liable to become so complex as to defeat its own ends, as well as being sensitive to human error.

Clearly, because of its speed of operation, and its capacity for storing and having rapid access to large quantities of information, a computer can play a vital role in making progress with the problem of controlling production. Not only can a computer produce rapid modifications to the original plan almost as soon as their need has been recognised but it can also issue instructions to the personnel who have to carry it out. It can also monitor the performance so that deviations, shortages and delays are pointed out quickly and can be dealt with. This function implies that some kind of feed-back for the computer has been organised.

Although we have here an extremely powerful tool at our disposal, long and careful study is necessary before it can be applied with confidence. No standard procedure is possible, because no two production units are identical. As we shall stress in the chapter on programming, computers have no common sense, and the story of the unit in the motor industry which produced tens of thousands of starting handles, while no doubt apocryphal, illustrates the need for careful planning. This fact no doubt accounts for the slow development of this important application of computers, but it will grow rapidly.

Automatic Control of Machine Tools

Besides carrying out calculations for scientific or commercial purposes from input on paper tape or punched cards, and giving the output on a printer, computers can be used to work in 'real time'. This expression means that the output of the computer program is used to control a physical process while it is in progress; for instance, a computer can have fed into it continuously measures of the performance of a steel rolling mill (the thickness of the steel output, temperature of the molten steel, pressure on the rollers, etc.), and from these figures it can be used to correct or optimise the performance of the steel rolling mill to give the output desired (e.g. steel of a certain defined thickness) by controlling, say, the speed of the rollers.

Such use of computers as the control or nerve centre of a manufacturing process is one of the aspects of automation, called Process Control.

Special computers can be connected to special machine tools to carry out a long sequence of machining operations under computer control. For instance, a computer-controlled machine tool can be fed with a paper tape, the holes in which instruct the machine tool what to do. The machine tool may have controls to move the material that is being machined precisely in any one of the three axes (x, y, z), control the tool needed (e.g. a mill, drill, tap, etc.) and control its rate of revolution. By this means one machine may produce very quickly and very accurately a complicated machined part. By conventional machines such a part would pass through a sequence of operations on various different machines, taking much longer and introducing a far greater chance of error.

Type Composition

Another kind of device which can be controlled by paper tape, and indeed by magnetic tape, is a typesetting machine. This fact is significant to large organisations which have computer-processed information to be communicated to the outside world. A good example is Post Office Telephone Directories. All the information to be printed in a Directory is contained in the billing files of the Post Office. Now this information can be read by a type-setting program, made up into columns and pages; typographical style applied, e.g. bold words, dots across the line, etc.; and then written to a magnetic tape in a form acceptable to the type-setting machine.

Above are listed just a few of the more interesting applications which computers are undertaking. Obviously there are many more applications which one could cover in this era when the computer is really flexing its muscles following its now universal acceptance as a fast, accurate, viable maid-of-all-work.

8 Programming

The first thing to emphasise is that computers do not, and cannot, solve problems. They merely carry out procedures given to them by the programmer. If by blindly carrying out the procedures in the correct order the result is a solution of the problem, so much the better.

The procedures are limited, and normally are very simple and rudimentary. They are built into the machine by the designer, who has to work to a limit of cost. We have already seen in Chapter 4 examples of the kinds of procedure which can be used.

Given a problem, then, the first step for the programmer is to decide in a general way how he proposes to obtain a solution. This means that the problem must be defined clearly. In some applications this may be easy enough; in others it may require months of close study.

The next step is to work out a detailed line of attack. In some cases the mathematics involved may consist only of some simple arithmetical operations. This is so in a number of commercial problems. But if the general approach is to reduce some real-life problem to a mathematical formulation it is necessary to decide which particular method we intend to use to obtain a solution. In some cases there is no known method, and one has to be developed specially. This can take time.

We must then reduce the mathematical procedures to a series of arithmetical operations. To take a simple example, if we had to extract a square root we should have to reduce this to a series of arithmetical steps. Or, in the case of an integration, we might have to evaluate it by some method such as Simpson's Rule.

This type of activity is known as Numerical Analysis. It is really a subject in its own right.

Programmers who are employed on this type of problem do require some mathematical training, but this is not necessary to the same degree in a vast range of other programming applications.

Whatever the application, however, a very careful and precise plan of operations is required.

Most people set out their plan in the form of a

block diagram. This sets out the various steps in the correct order but does not contain much detail. It does show how a process can be repeated if required and also shows various alternative procedures, the adoption of which may depend on the previous results.

The next step is to write out in terms of the machine functions the many individual steps in the correct order which are specified broadly in the block diagram. There may be thousands of these elementary instructions. These have to be coded in machine code before they can be fed into the computer.

The program must then be tested. It is virtually impossible to produce a program free of errors. Any error will cause something odd to happen, and is usually easily detected when the program is run on the computer. It may take longer to find why the error occurred and correct it.

Finally, we add the raw data for the computation and run the program. The raw data could, for example, be the constants in some mathematical problem or the details of individual men in a payroll calculation.

At the risk of being repetitive, we must again stress the paramount importance of planning the program with extreme care and accuracy. The computer has the qualities of very high speed and also accuracy, but it has absolutely no common sense. It cannot decide what to do and has no appreciation of the results.

Once you start to run a program, you have for the time being lost control. Everything must be thought of in advance. If at some stage a choice has to be made, you must have foreseen this and specified on what basis the choice is to be made; and given explicit instructions.

It will be obvious that programming carried out in this fashion can be a laborious process.

Sub-routines

As computers developed, certain features of programming became apparent. In mathematical applications, for example, it frequently happened that the same sequence of instructions could be repeated several times in one program, and could crop up again and again in different programs. For example, it might be a question of calculating the value of sine x given x. It was an obvious step to store the sequence of program steps to do this, so that it could be referred to whenever required and inserted in the correct place in the complete program.

Naturally a library of such sub-routines as they came to be called was soon built up for each particular computer. The next step was to identify each of these sub-routines by a code number and to add to the machine functions an order which could call up the various sub-routines required.

The range of problems capable of being dealt with in this fashion was, of course, limited by the sub-routines available.

It would have been a great convenience if it had proved possible to write a program which consisted entirely of orders which called up successive sub-routines. Attempts were made to do this, but soon a better idea cropped up. This was the development of Autocodes.

This involved the development of a special kind of program called a Compiler. This caused the computer itself, when so instructed, to produce a program in machine code, the instructions used being limited to a fairly brief set—known as the source language.

Provided a compiler is available, the writing of a program in autocode is much faster than having to write the full program in machine language.

An immediate development was to write the source language codes in alphabetical form instead of in numbers. (Since alphabetical characters are coded in binary form in a computer, this presents no complication.) Thus the computer can accept the instruction 'DEDUCT TAX' as easily as an instruction 34768.

It was now possible to make use of mnemonic words in the autocode, and this made programming easier to learn and to understand when checking.

Several autocodes were developed. As a result of agreement between programmers in various countries, there are now a limited number of generally accepted autocodes.

FORTRAN is the code used largely in mathematical work.
ALGOL is similarly used.
COBOL is a business oriented programming language.

Each different computer needs a compiler for each of the autocodes. In some cases, where a range of compatible computers is produced, the smaller versions have compilers which can only accept a sub-set of the instructions on ALGOL etc.

Recent application of computers to the problems of numerical control of machine tools—an up-to-date form of automation—have led to the development of another specialised symbolic programming language known at APT (Automatically Programmed Tools). Here, a limited vocabulary in symbolic language enables surfaces to be defined, tools specified, and tool positions and orientations relative to as many as five axes to be controlled.

All this is, of course, translated into detailed instructions on the paper tape which controls the machine.

It is likely that the system will be developed further by means of a visual display. This could be presented on a cathode-ray tube. As the program develops it may be possible to show images on the tube which will simulate the progress of the part being machined. This will enable alterations to be made if necessary before committing the tool to the real job.

There are several specialised types of program, some of which should be mentioned here.

So-called diagnostic programs are produced for the benefit of the servicing engineer. They can be run through to pin-point areas of failure in the machine. This type of program is produced in the development laboratories.

The increasing use of time sharing in large computers required the development of special programs which in effect control the internal organisation of the machine and allocate time between the various peripheral units, as well as keeping the whole complicated process tidy; these are known as 'Operating Systems'. This type of program is also closely related to the design of the computer and is produced in the development laboratories.

Another problem which is being tackled is a kind of tidying up process, in that attempts are being made to translate one program into another. We may have a program written in some autocode and we wish to translate it so that it can then be used on another machine.

This problem can be particulary difficult if we are talking of machines made by different manufacturers, but even in machines made by the same manufacturer it is sometimes far from easy. Differences in word length can raise serious difficulties; some instructions may refer to hardware which has no counterpart in the second machine; occasionally instructions make sense when dealt with in sequence but nonsense if considered individually.

In the future this problem will become easier to

handle. There is a tendency for symbolic languages to become more machine independent, and the differences between commercial and mathematical computers are tending to disappear.

Software

In the computer world 'jargonese' has rapidly sprung up, thc two most common terms being 'hardware' and 'software'.

> *Hardware* is all the physical components of a computor system, covering processors, peripherals, storage devices, etc.
> *Software* is all the programs and sub-routines required to drive the hardware.

In these days of computer proliferation the technology used in hardware is very much the same, regardless of manufacturer. Obviously there are differences, but, when compared with the capital sum involved in purchase, these differences are rarely significant. Thus, in order to attract orders, computer manufacturers had to offer something more. This is software. Many users are performing the same applications on their machines, and manufacturers have come to realise that, by programming these standard applications and offering the programs with their machines, significant savings could accrue to the user.

However, it was realised that, while some applications conformed to a standard set of national or international rules (e.g. Payroll), other applications performed a common set of functions in a way peculiar to individual users. Thus manufacturers' software developed in two ways.

Packages

A software package is designed to cope with the first-mentioned application type.

That is, where the whole operation conforms to nationally or internationally accepted standards and the whole process can be programmed by an independent body with no regard to the specific company which is to use the system. Therefore all the processing is bound up inside the package, and the user inputs data and accepts the output produced.

Payroll is a good example of this, whereby the rates of tax, insurance, etc., are nationally specified. Thus, the user inputs details of hours worked, pay rates and special deductions; then the package can perform gross-net calculations and print out payslips; credit transfers and payroll analyses. Payroll has the further advantage that coin denominations are also standard nationally; thus each pay packet can be broken down into the individual notes and coins which it will contain, and when these are totalled a coin analysis is produced for the bank. The person or organisation responsible for making up pay packets presents this analysis to the bank, draws precisely the correct amount of coins and notes (in quantity as well as value) and proceeds to make up the pay packets.

Modular Software

This software deals with that type of application where several. common functions have to be performed but the sequence and way in which they are used is peculiar to the company using the system.

An example of this is computer type composition. The basic problems are to form text into lines, apply hyphenation decisions, proof and correct the text, and then drive one or more of a series of type-setting machines to produce the printed word..

Further complications are added by the extreme flexibility offered to authors by printers in such things as page size, typefaces, etc.

The basic functions can be programmed by an independent body; but the final decisions on how the information is used must remain with the printer. Computer manufacturers therefore supply a series of program modules (like sub-routines) which perform basic functions but need to be linked into a system by the user's own programmers.

9 Communications Systems

The Growing Need

As more and more people, departments and companies become dependent upon the computer and its capabilities, the need to provide an easy and cost-effective method of distributing this mathematical and storing capability becomes increasingly evident. In the past there were two main restrictions in the provision of computing power to the ultimate user: geographical location, and the programming requirement to extract the information once the problem of distance had been overcome. These restrictions resulted in the user preparing a special type of input document suitable only for the computer, the sending of that document to the computer department, and the final receipt of computer-printed output after, perhaps, a number of phone calls between the user and the computer department concerning the accuracy of the initial input information. This process could take anything between half a day, if the user was in the room next door to the computer, to a whole week if the input and output documents had to be posted. Obviously, with the increased need to process and store information on an ever decreasing time scale the method outlined above was just not suitable. Faster methods of document handling and easier access to the machine were required, and an easier language with which to communicate with the computer became essential.

Centralisation and Decentralisation

In trying to solve the problem outlined above, the users of computers began to question whether the provision of a centralised computing service was the right choice. Perhaps the problems of geographical distance and the differing needs of users could be resolved by the installation of smaller computers, tailored to the users' specific needs, based at his location; a distributed computing philosophy. Whilst this approach could possibly satisfy all that is required from a user's point of view, the cost of providing such a system could in some cases become excessive.

The general rule applying here is that the computing power bought per pound increases geometrically, not arithmetically; this therefore favours the large central Data Processing (EDP) organisation. To overcome these problems of cost and ultimate user control, a third approach has been taken which goes a long way towards satisfying both requirements, namely data communications. It is this method of communicating data over existing telegraph and telephone lines that concerns us in this chapter.

By providing the user with the equipment, programs and operating systems, a user can now ring up the computer directly, call up his own programs, enter data, process and receive the results back on his own printer in a matter of minutes. Further on in this chapter we shall be looking at the characteristics of this equipment, known as Terminals; but before we are able to do this there are a few points of 'jargon' that need to be explained, and some overall considerations that need to be examined in broad detail.

Jargon

Between the Terminal and the computer there

must obviously be some connection over which the data can be transmitted. When the terminal is located close to the computer, up to, say, 500 yards and on the same premises, then a form of local cabling can often be used. However, when the terminal is a long way from the computer, then suitable communication lines have to be provided. In the U.K. these are supplied by the General Post Office. A number of services are available and depend upon the type of line being used. When used for the transmission of data, a line is graded according to how many bits of data information it can transmit in one second; therefore a 600 bit per second line is capable of transmitting sixty data characters every second if each character is comprised of ten individual data bits. The following line speeds are generally available within the U.K.: 50, 75, 100, 110, 200, 300, 600, 1200 and 2400 bits per second. Speeds below 200 b.p.s. are normally available over telegraph lines, whilst the faster speeds are available over telephone lines. A telegraph line transmits the bits as a voltage signed–i.e. a 1-bit represented by +80 V and a 0-bit by –80 V. Telephone lines, however, are quite different and use a method of modulation to transmit the data. As this latter method does not involve the switching of fairly large voltages, with the attendent problem of equipment reliability, very high data transmission (carrying) rates can be attained. To achieve the necessary modulation and demodulation processes a piece of equipment known as a Modem (MOdulator/DEModulator) has to be connected to each end of the telephone line. The signals from the terminal are presented to the modem, where they are modulated on to a carrier wave and transmitted down the line. At the computer end the modulated signal is demodulated by the receiving modem and the resultant signal, identical to that transmitted by the terminal, is presented to the computer.

A second aspect of the transmission medium is the manner in which it is to be used. If data is to be transmitted in one direction only, then this is known as Simplex Transmission, the opposite to this being Full Duplex transmission, which is the transmission of data in both directions simultaneously.

There is a stage known as Half Duplex where data is able to be transmitted in both directions but not at the same time. At present most terminals use this latter method of transmission.

The last point that needs to be discussed here is

that of transmission errors. Because the data is being transmitted in a different form from that used by local computer peripherals, it is necessary that some form of data checking be performed so that any errors occuring during transmission may be identified. This is usually achieved by adding parity bits to the data characters by the transmitting terminal. The receiving terminal performs a parity checking function and compares the received characters with the one it has built up—any errors at this stage will cause an appropriate error procedure to be initiated. When this condition exists one of the popular methods of correcting the error is for the terminals to initiate automatically a re-transmission of the data last transmitted, this process being repeated until the data has either arrived correctly or a predetermined number of repeats has been reached, and the transmission is terminated with an indication to the terminal operator to this effect.

On-line and Off-line Data Transmission

When considering the methods by which data may be communicated between one location and the computer centre, we must consider whether the terminal at the remote location need be physically connected to the computer or whether an intermediate stage can be of advantage; that is, the originating terminal sends the data to another terminal at the computer centre. The second terminal would produce the data in a suitable form for subsequent input to the computer. This method of transmission is known as off-line working and the most common data media is paper tape, although there has recently been an increasing tendency to

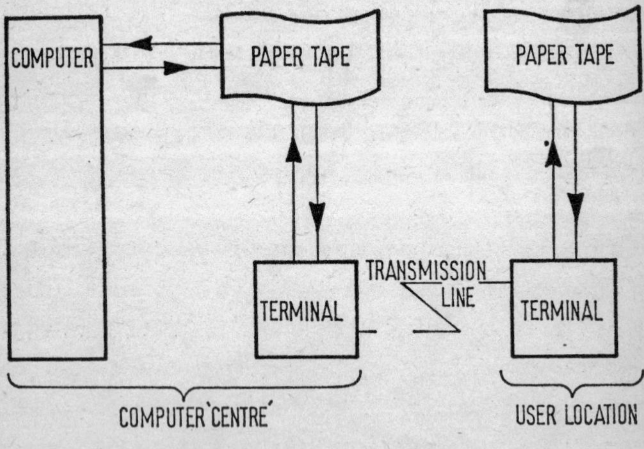

FIG 9.1

use magnetic tape. Fig. 9.1 illustrates the principles involved in off-line data transmission.

This method of data transmission is very effective when it is necessary to ensure that data gets transmitted rapidly and where the terminal user does not require immediate access to the computer functions. However, when the user's application does require access to the mathematical and storage functions, the terminal must be connected to the computer. This is achieved by attaching the transmission line directly to the computer, normally (though not necessarily) through a control unit that is capable of supporting a number of lines and terminals. Fig. 9.2 indicates how this is achieved.

Terminals

There are, as one would expect, a large number of different terminals available to satisfy the differing needs of the terminal user, and as time goes by an increasing variety are becoming available. We shall now look at some of the most popular terminals currently available and shall see that each has its own particular characteristics that render it useful for the now identified application areas.

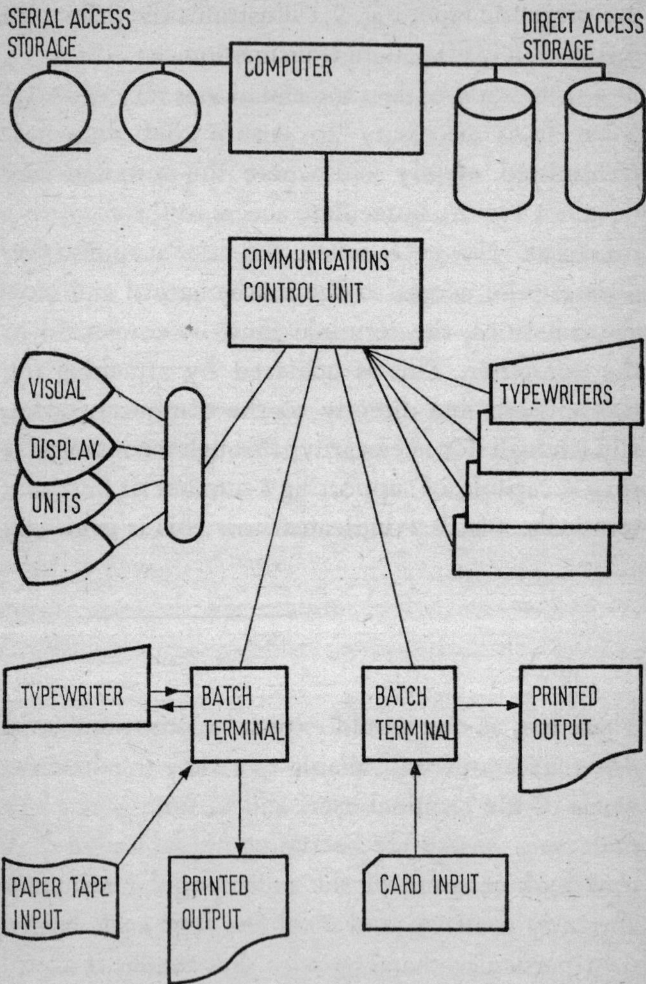
SERIAL ACCESS STORAGE
COMPUTER
DIRECT ACCESS STORAGE
COMMUNICATIONS CONTROL UNIT
VISUAL
DISPLAY
UNITS
TYPEWRITERS
TYPEWRITER
BATCH TERMINAL
BATCH TERMINAL
PRINTED OUTPUT
PAPER TAPE INPUT
PRINTED OUTPUT
CARD INPUT

FIG 9.2

Typewriters

These terminals have characteristics very similar to those of the normal office typewriter; that is, a keyboard of similar layout but with some control keys and a printer mechanism for printing the keyed information and reply information from the computer. Every time a key is pressed the character is printed and, at the same time, is encoded by the terminal and transmitted down the line to the computer; at the end of the message, a control key is pressed and the resulting character, when received by the computer, informs the program that the message is complete and may therefore be processed. After processing by program the answer is transmitted character by character back to the terminal and printed out. The operator then types in the next message. Thus a conversation between the terminal operator and the processor is in progress and is often referred to as interactive (conversational) working. An example of interactive working is shown below:

```
line 1 ↑   LET A = 21·25
line 2 ↑   LET B = 16·791
line 3 ↑   TYPE A * B
line 4           A * B    = 356·80875
line 5 ↑
```

In this example the operator is invited to type by the '↑' sign sent from the computer. He keys in the first line and presses a control character to indicate end of message (this character is not printed).

The computer program replies with another invitation to type as shown in line 2. In line 3 the operator has keyed the word TYPE before his message, which is an instruction to the computer to do a certain function, and after transmission of that line the computer replies with the answer:

$$21{\cdot}25 \times 16{\cdot}791 = 356{\cdot}80875$$

followed by a further invitation to type on line 5. Because the operator usually spends a lot of time thinking about the next line and typing at a fairly slow rate, it is not necessary to have a very fast and therefore expensive terminal; thus the normal speed at which this type of device operates is around ten characters per second.

Visual Display Units

The visual display unit is another terminal that usually works in the conversational mode; but, instead of having printed output, it uses a cathode ray tube to display the information. In some ways it can be likened to a television screen showing only sub-titles. These terminals, capable of operating at speeds of up to 480 characters every second, are being extensively used for displaying information as a result of an enquiry initiated by the operator. For example, this type of terminal is in use for airline reservation systems, where the operator can key in a customer's flight requirement and immediately receive on the display flight details that conform to the initial requirements. Should it be necessary for the display information to be copied, then this is achieved by attaching a printing mechanism to the visual display terminal.

Terminals of this type are usually designed to sit on the top of a desk and have the ability to display up to 2000 characters of information.

Batch Terminals

Batch terminals, as their name implies, are designed for the transmission and reception of data in bulk form; that is, the direct opposite to that of interactive terminals. These terminals are normally available in a variety of forms and may include a small processor to enable a limited amount of pre-processing to be performed before transmitting the data to the central computer—e.g. check digit verification. A typical terminal would be equipped with a card reader, line printer and console typewriter, the latter peripheral being used for typing instructions to or from the processor (e.g. the computer informing the operator that the line printer is running out of paper). Such a terminal appears to the user as virtually his own computer, in that he can prepare his data locally, control the processing required and receive printed output without moving from his location. This method of operation, known as Remote Job Entry, is becoming increasingly important. Typical card reading speeds are 300 cards per minute and printing at 600 lines per minute; these speeds, however, are very dependent upon the speed of line used for transmission. The most popular speeds for batch terminals are 1200 and 2400 bits per second.

The Future

The increasing interest being shown in data communications is going to result in even faster and more cost-effective terminals being used in ever increasing application areas. The now common speed of 2400 bits per second will increase to 9600 bits per second, enabling card and line printing speeds of up to 600 cards and 1000 lines per minute respectively. Visual display units will be used extensively in such areas as order processing, stock recording, production processing and management information systems.

Some of the larger communications users, however, are beginning to establish that conversational terminals may not be the complete answer. To clarify this, consider the case of a Building Contractor having centralised costing and scheduling systems controlling a series of sites in various locations on a national basis. Each of the sites will employ staff according to local rules in force regarding terms of employment and payment. Thus, although each site requires large high-powered computing for its construction schedules, and needs to supply its cost and budgetary information to the central accounting function, there is a case for producing the payroll locally. The ideal requirement for such a situation is some form of computing power on site which can also be connected direct to the central machine. In response to this kind of demand a new range of 'mini-computers' is evolving. These are small cheap processors with restricted backing store and software facilities which can therefore be used as terminals to any kind of large main-frame computer. They will perform basic accounting and scientific functions locally at very low cost and also to be switched into the central machine, becoming a terminal for access to greater computer power and head office functions.

It is anticipated that this form of terminal equipment will become larger and more sophisticated, yet will remain within the cost constraints imposed. One can see that large organisations, such as the Common Market, will generate a need for common computor languages and inter-governmental links. Perhaps the result will be that, in the not too distant future, by using the commonalities of their computers 'Nation shall speak peace unto Nation'.

TEACH YOURSELF BOOKS

MARKETING

John Stapleton

One of the better definitions of marketing is that—'Marketing is the management process which identifies, anticipates and supplies customer requirements efficiently and profitably.' This is not new as a basic concept but what is new is the development of modern sophisticated techniques being increasingly introduced into the different areas of marketing.

In this book, John Stapleton examines these techniques in turn and demonstrates how they act together as components of the total marketing activity. By verbalising the concepts of marketing and by clarifying the reasoning behind marketing practice, the author provides an analysis of these marketing components, their function and relevance.

The book also provides a basic but wide description of marketing activity, with individual chapters covering specific areas such as marketing research, product planning and pricing, marketing planning, distribution and selling, and the sales force. Professional marketing men will find here a valuable source of information and stimulus, while students of the Institute of Marketing's Qualifying Certificate and Diploma courses will find this book an ideal introductory text.

UNITED KINGDOM	85p
AUSTRALIA	$2.45*
NEW ZEALAND	$2.45
CANADA	$3.50

*recommended but not obligatory

ISBN 0 340 05967 2

TEACH YOURSELF BOOKS

COMPUTER PROGRAMMING/FORTRAN

A. S. Radford

A modern digital computer is capable of performing over one million operations a second, each of which it has to be instructed to perform. The form in which instructions are given to the computer is called a program and this must be written in a code or language intelligible to the computer.

This book provides a basic course in Fortran, a computer language widely adopted for scientific and general purpose commercial use. It begins with an outline of the development of the digital computer and computer languages and by describing the technique of flow charts—the means of establishing the sequence of events to be performed by the computer. The remainder of the book is concerned with explaining, analysing and teaching Fortran language, covering a wide variety of programs and operations. Exercises to develop the programmer's proficiency in Fortran are incorporated in the text, which has been realistically geared to the needs of the self-educator as well as to those of the computer science student.

UNITED KINGDOM	95p
AUSTRALIA	$3.05*
NEW ZEALAND	$3.20
CANADA	$3.50

*recommended but not obligatory

ISBN 0 340 19495 2

TEACH YOURSELF BOOKS

OPERATIONAL RESEARCH

M. S. Makower & E. Williamson

Operational Research is the application of scientific method to management decision-making. It is being used to tackle a wide variety of problems, ranging from space-exploration to the control of supermarket stocks.

This book introduces and discusses some of the more important techniques which have been developed to help solve such problems. These include forecasting methods, the theory of queues, linear programming, and network analysis. Explanation and illustration of these techniques is carried out through worked examples, and each chapter ends with further exercises (and solutions) to enable the reader to test his understanding of the methods described.

Operational Research is a fascinating and increasingly topical subject. Those with a slight taste for numbers and at least some recollection of school mathematics should follow with ease the ideas presented.

An introduction to OR for business and management students and for all those involved in forecasting, planning, and decision-making.

UNITED KINGDOM	£1.25
AUSTRALIA	$3.95*
NEW ZEALAND	$4.15
CANADA	$4.50

*recommended but not obligatory

ISBN 0 340 20232 7